PHOTONICS
An Introduction

P.R. Sasi Kumar

Department of Physics
Maharaja's College
Ernakulam, Kerala

PHI Learning Private Limited

New Delhi-110001
2012

₹ 225.00

PHOTONICS: An Introduction
P.R. Sasi Kumar

© 2012 by PHI Learning Private Limited, New Delhi. All rights reserved. No part of this book may be reproduced in any form, by mimeograph or any other means, without permission in writing from the publisher.

ISBN-978-81-203-4532-4

The export rights of this book are vested solely with the publisher.

Published by Asoke K. Ghosh, PHI Learning Private Limited, M-97, Connaught Circus, New Delhi-110001 and Printed by Raj Press, New Delhi-110012.

Contents

Preface   *xi*
List of Units   *xiii*

1. Radiometry   1-9
1.1 Introduction   *1*
1.2 Radiant Power (Radiant Flux ϕ)   *2*
1.3 Radiant Energy (Q)   *2*
1.4 Units Related to the Transmitter   *2*
    1.4.1 Radiant Emittance (W)   *2*
    1.4.2 Radiant Intensity (I)   *3*
    1.4.3 Radiance (L)   *3*
1.5 Units Related to the Receiver   *4*
    1.5.1 Irradiance (E)   *4*
    1.5.2 Point Source   *4*
1.6 Lambert's Law   *5*
1.7 Extended Source   *6*
1.8 Diffuse Reflector   *6*
1.9 Total Radiant Power from a Surface   *6*
Review Questions   *9*

2. Elements of Optical Phenomena in Semiconductors   10-32
2.1 Semiconductor Physics   *10*
2.2 Energy Distribution Function   *11*
2.3 Density of States   *12*
2.4 Effect of Doping   *13*
2.5 Quasi-Fermi Levels under Charge Injection   *14*
2.6 Band Structure   *16*
2.7 Direct Band Gap Semiconductors   *16*
2.8 Indirect Band Gap Semiconductors   *17*
2.9 Electron Hole Pair Creation and Recombination   *18*
2.10 Recombination Mechanisms   *19*
2.11 Radiative Recombination Efficiency   *19*

2.12 Absorption Mechanisms in Semiconductors 21
    2.12.1 Band to Band Transition 21
    2.12.2 Indirect Transition 22
    2.12.3 Low Energy Absorption 22
2.13 Excitons 24
2.14 Emission in Semiconductors 25
    2.14.1 Direct Band Transition 26
    2.14.2 Indirect Band Transition 26
    2.14.3 Impurity Band Emission 26
2.15 Franz-Keldysh Effect 28
2.16 Stark Effect 28
2.17 Stokes Shift in Semiconductor 28
2.18 Auger Recombination 29

Review Questions 32

3. Light Emitting Diode   33–48

3.1 Introduction 33
3.2 Luminescence 33
    3.2.1 Photoluminescence 34
    3.2.2 Cathodoluminescence 34
    3.2.3 Electroluminescence 34
3.3 Electroluminescence in *p-n* Junction 34
3.4 Radiative Recombination in LED 36
3.5 LED Drive Circuit 36
3.6 LED Characteristics 37
    3.6.1 Spectral Response 37
    3.6.2 Current-Light Power Characteristics 38
    3.6.3 *I-V* Characteristics 39
    3.6.4 Output Power-Time Characteristics 39
    3.6.5 Temperature Dependence on Light Emission 40
    3.6.6 Temporal Response 40
3.7 Light Output from LED and Optical Responsivity 41
3.8 Efficiency of LED 42
    3.8.1 Injection Efficiency 42
    3.8.2 Radiative Recombination Efficiency 43
    3.8.3 Extraction Efficiency 43
3.9 LED Material Choice 44
3.10 LED Construction 44
3.11 Planer Surface Emitting LED 45
3.12 Double Hetrojunction (DH) LED Structure 45
3.13 Edge Emitting LED 46

Review Questions 47

4. Semiconductor Lasers   49–57

4.1 Introduction 49
4.2 Difference between LED and SDL Light Emission 49
4.3 Laser Action in a *p-n* Junction 50

4.4   Conditions for Laser Action   *50*
   4.5   Threshold Gain   *51*
   4.6   Power Output from a SDL   *52*
   4.7   Basic Structure of a Junction Laser   *52*
   4.8   Hetrojunction Lasers   *54*
         4.8.1   Single Hetrostructure (SH) Laser   *54*
         4.8.2   Double Hetrostructure (DH) Laser   *54*
         4.8.3   Stripe Geometry Hetrostructure   *55*
   4.9   Quantum Well Lasers   *55*
   4.10 Distributed Feedback Laser   *56*
   Review Questions   *57*

5. Photodetectors   58–83
   5.1   Introduction   *58*
   5.2   Photoconductive Detectors   *58*
   5.3   Photoconductive Materials   *60*
         5.3.1   Doped Semiconductors   *60*
         5.3.2   Cadmium Sulphide and Cadmium Selenide (CdS and CdSe)   *60*
         5.3.3   Lead Sulphide (PbS)   *61*
         5.3.4   Indium Antimonide (InSb)   *61*
         5.3.5   Mercury Cadmium Telluride (HgCdTe)   *61*
   5.4   Junction Photodiode   *61*
         5.4.1   *p-n* Junction   *61*
         5.4.2   Photodiode: Principle   *62*
         5.4.3   *V-I* Characteristics of a Photodiode   *62*
         5.4.4   Design of Photodiodes   *63*
         5.4.5   Equivalent Circuit of a Photodiode   *63*
   5.5   PIN Photodiode   *65*
         5.5.1   Expression for Photocurrent   *65*
         5.5.2   Response Time   *67*
   5.6   Avalanche Photodiode   *68*
   5.7   Modulated Barrier Photodiode   *70*
   5.8   Metal Semiconductor Photodiode (Schottky Barrier Photodiode)   *71*
   5.9   Phototransistor   *71*
   5.10 Microcavity Photodiode   *73*
   5.11 Photomultiplier Tube   *74*
         5.11.1   Photoelectric Effect   *74*
         5.11.2   Photomultiplier Tube (PMT)   *75*
         5.11.3   Noises in PMT   *76*
   Review Questions   *82*

6. Solar Cells   84–95
   6.1   Introduction   *84*
   6.2   Principle of Operation   *84*
   6.3   Spectral Response   *88*

- 6.4 Homojunction Solar Cells *89*
- 6.5 Hetrojunction Solar Cells *89*
- 6.6 AlGaAs/GaAs Hetrojunction Solar Cells *90*
- 6.7 Thin Film Solar Cells *90*
- 6.8 Schottky–Barrier Solar Cells *91*
- 6.9 Cascade Hetrojunction Solar Cells *92*
- 6.10 Material Requirements *92*
- 6.11 Temperature and Radiation Effects on Solar Cell Performance *92*
- 6.12 Optical Concentration *93*
- 6.13 Device Fabrications *93*

Review Questions 95

7. Fibre Optics   96–121
- 7.1 Optical Fibre *96*
- 7.2 Total Internal Reflection *97*
- 7.3 Types of Optical Fibres *98*
    - 7.3.1 Step Index Fibre *99*
    - 7.3.2 Graded Index Fibre *100*
- 7.4 Rays in an Optical Fibre *101*
- 7.5 Numerical Aperture *101*
- 7.6 Type of Modes *102*
    - 7.6.1 Guided Modes *103*
    - 7.6.2 Radiation Modes *103*
    - 7.6.3 Leaky Modes *103*
- 7.7 *V*-Parameter *103*
- 7.8 Multimode and Single Mode Fibres *104*
- 7.9 Coherent Optical Fibre Bundle *105*
- 7.10 Optical Fibre Cables *106*
- 7.11 Optical Fibre Materials *107*
- 7.12 Glass Fibres *108*
- 7.13 Plastic Fibres *108*
- 7.14 Rare Earths Doped Optical Fibres *109*
- 7.15 Infra Red Fibres *109*
- 7.16 Zero Dispersion Fibre *109*
- 7.17 Polarization Maintaining Fibres *110*
- 7.18 Attenuation in Optical Fibres *110*
    - 7.18.1 Absorption Loss *111*
    - 7.18.2 Scattering Losses *111*
    - 7.18.3 Dispersion Mechanisms *112*
    - 7.18.4 Bending Losses *113*
    - 7.18.5 Interface Inhomogeneities *113*
    - 7.18.6 Joining Losses *113*
- 7.19 Fibre Connectors, Splices and Couplers *114*
- 7.20 Fabrication of Optical Fibres *114*
    - 7.20.1 Direct-melt Double-crucible Method *114*

7.21 Fibre Drawing Process *115*
7.22 Optical Fibre Communication System *116*
7.23 Wavelength Division Multiplexer *117*
7.24 Advantages of Optical Fibre Communication System *118*
7.25 Fibre Optic Sensors *118*
Review Questions *120*

8. Modulation of Light   122–134
8.1 Introduction *122*
8.2 Direct and External Modulations *122*
  8.2.1 Direct Modulation *122*
  8.2.2 External Modulation *123*
8.3 Direct Modulation of LED *123*
8.4 Direct Modulation in SDL *124*
  8.4.1 Extrinsic Factors *125*
  8.4.2 Intrinsic Factors *125*
8.5 External Modulators *125*
  8.5.1 Electro Optic Effect *126*
  8.5.2 Pockel Effect in KDP Crystal *126*
8.6 Principle of Phase Modulation (Electro Optic Retardation) *127*
8.7 Electro Optic Amplitude Modulator *128*
8.8 Kerr Modulators *130*
8.9 Magneto Optic Modulation *130*
8.10 Acousto Optic Effect *130*
Review Questions *134*

9. Non-linear Optical Processes   135–140
9.1 Introduction *135*
9.2 Second Harmonic Generation *136*
9.3 Self-focussing and Defocussing *138*
9.4 Optical Parametric Interactions *139*
9.5 Four Wave Mixing *139*
9.6 Multiphoton Absorption *140*
Review Questions *140*

10. Integrated Optics   141–158
10.1 Introduction *141*
10.2 Waveguide Structure *142*
10.3 Waveguide Devices *143*
  10.3.1 Passive Waveguide Devices *143*
  10.3.2 Active Waveguide Devices *149*
10.4 Phase Modulator *150*
10.5 Mach–Zehnder Interferometric Modulator *151*
10.6 Directional Coupler Switch *151*

10.7 Active Waveguide Devices Based on Acousto Optic Effect *153*
  10.7.1 Acousto Optic Bragg Modulator *153*
  10.7.2 Acousto Optic Spectrum Analyzer *154*
10.8 Active Waveguide Devices Based on Magneto Optic Effect *154*
10.9 Active Waveguide Devices Based on Thermo Optic Effect *155*
10.10 Active Waveguide Devices Based on Optical Non-linearity (Optical Bistable Device) *156*
10.11 Waveguide Coupling *156*
10.12 Opto Electronic Integrated Circuit *157*
Review Questions *158*

11. Holography   159–166
11.1 Introduction *159*
11.2 Method of Holography *159*
  11.2.1 Recording of Hologram *160*
  11.2.2 Reconstructing of Image *160*
11.3 Conditions for Recording Holograms *161*
11.4 Hologram Recording Materials *161*
11.5 Principle of Holography *162*
11.6 Types of Holograms *163*
  11.6.1 Volume Holograms *163*
  11.6.2 Rainbow Holography *163*
  11.6.3 Colour Holography *163*
11.7 Characteristics of a Hologram *164*
11.8 Applications of Holography *165*
Review Questions *166*

12. Display Devices   167–173
12.1 Introduction *167*
12.2 Cathode Ray Tube (CRT) *167*
12.3 Liquid Crystals *169*
12.4 Liquid Crystal Display (LCD) *171*
12.5 Charge Coupled Devices (CCD) *172*
12.6 Plasma Display *173*
Review Questions *173*

13. Lasers   174–192
13.1 Introduction *174*
13.2 Basic Principle of Laser Action *174*
13.3 Parts of a Laser System *177*
13.4 Properties of a Laser Beam *178*
13.5 Optical Resonator *178*
  13.5.1 Parallel Plane Resonator *179*
  13.5.2 Confocal Resonator *179*

        13.5.3 Spherical Resonator   *179*
        13.5.4 Large Radius Resonator   *179*
        13.5.5 Hemispherical Resonator   *180*
    13.6 Modes of a Planar Optical Resonator   *180*
    13.7 Quality Factor of an Optical Resonator   *181*
    13.8 Three Level Laser System   *182*
    13.9 Four Level Laser System   *183*
    13.10 Ruby Laser   *183*
    13.11 Nd:YAG Laser   *184*
    13.12 He–Ne Laser   *185*
    13.13 CO_2 Laser   *187*
    13.14 Dye Laser   *188*
    13.15 Semiconductor Laser   *189*
    13.16 Applications of Lasers   *190*
    Review Questions   *192*

14. Advances in Photonics   193–204

    14.1 Raman Scattering   *193*
    14.2 Photorefractive Effect   *194*
    14.3 Optogalvanic Effect   *196*
    14.4 Photothermal Deflection Effect   *197*
    14.5 Photorefraction in a Diffusing Medium   *198*
    14.6 Squeezed State   *199*
    14.7 Optical Solitons   *200*
    14.8 Optical Bistability   *200*
    14.9 Optical Interconnect   *202*
    14.10 Photonic Switches   *202*
    14.11 Optical Computers   *203*
    14.12 Ultrafast Phenomena   *203*
    Review Questions   *204*

References   **205**

Index   ***207–210***

Preface

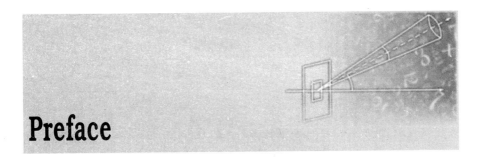

With the advent of lasers, technology associated with light called photonics (opto electronics) is fastly advancing. It involves tremendous applications in the fields of communication, science and technology, medicine, image processing, defence, optical computing and others. By considering the importance of this area many universities all over the world have introduced photonics as one of the subject in the curriculum. Many Indian universities have also adopted photonics at undergraduate and postgraduate level and also for engineering courses.

This book is meant for the undergraduate students of photonics as a textbook and undergraduate students of Physics as a reference book. Many of the undergraduate students are facing difficulty in understanding the subject while using the commonly available books in this field. The purpose of this book is to introduce the topic to the undergraduate students without using much mathematics.

Many topics are included in a simple way. It contains topics on radiometry, optical process in semiconductors, light emiting diodes, photodetectors, solar cells, fibre optics, non-linear optics, modulation, integrated optics holography, display devices and advancements made in Photonics. Each chapter is included with review questions and solved problems.

I acknowledge my thanks to the Directorate of Collegiate Education, Trivandrum, Maharaja's College, Ernakulam and Prof. VPN Nanpoori, CUSAT for the encouragement and support provided to write this book. I am also indebted to my wife Latha MS, and my children Meena and Ghanshyam, for their cooperation and patience while completing this project.

P.R. Sasi Kumar

List of Units

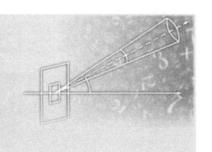

Symbol	Radiometric quantity and unit	Photometric quantity and unit
ϕ	Radiant power (W) flux	Luminous power (lm) flux
Q	Radiant energy (J)	Luminous energy (lms)
W	Radiant emittance (Wm^{-2})	Luminous emittance (lm m^{-2})
I	Radiant intensity (Wsr^{-1})	Luminous intensity (cd = lm sr^{-1})
L	Radiance (Wm^{-2} sr^{-1})	Luminance (cd m^{-2})
E	Irradiance (Wm^{-2})	Illuminance (lm m^{-2})

1

Radiometry

1.1
INTRODUCTION

Electromagnetic radiation in the wavelength range of the electromagnetic spectrum consisting of ultraviolet (UV), visible and infrared (IR) is considered as optical radiation (about 0.3 µm to 15 µm). As light is a form of energy, it is essential to use proper units to specify the physical quantities such as radiant energy, power etc. Radiometry is concerned with the measurement of the physical quantities of the electromagnetic radiation in the whole electromagnetic spectrum. Measurement of these quantities in the visible region of the optical spectrum is referred to as photometry. By conversion, radiometric units are denoted with subscripts e (energy) and the photometric units are denoted with the subscripts v (visible).

Figure 1.1 shows the natural response of the human eye to light at different wavelengths. These curves are called relative luminosity curves (or spectral responsivity) of the human eyes, which are obtained after a large number of measurements on observers to determine the relative effectiveness of monochromatic radiation to brightness sensation. The spectral responsivity of the eye falls to zero except in between 400 nm and 700 nm and is taken as unity at 555 nm, where the eye has maximum sensitivity.

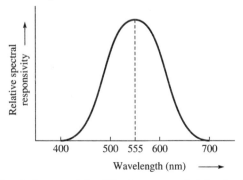

Figure 1.1 Spectral responsivity of the human eye (relative luminosity curve).

Photonics: An Introduction

The scale is conveniently set equal to unity at the maximum of the curve. By definition, at the peak sensitivity of the relative luminosity curve 1 watt of radiant power correspond to 685 lumens.

1.2
RADIANT POWER (RADIANT FLUX ϕ)

The total radiant energy, Q, emitted in all direction by a radiating source in 1 sec. is termed as radiant power.

$$\phi = \frac{Q}{t}$$

It is also equal to the rate of change of radiant energy.

$$\phi = \frac{dQ}{dt}$$

The unit of radiant power in the SI is watt (W). The corresponding photometric quantity is called luminous power (luminous flux). The unit of luminous power is lumen (lm).

1.3
RADIANT ENERGY (Q)

The radiant energy is given by

$$Q = \int_0^t \phi \, dt$$

If the radiant flux is uniform over the time t, then

$$Q = \phi t$$

The radiant energy is measured in joules (J). The photometric quantity is luminous energy, and has the unit lumen-second (lms).

Units used to specify the optical radiation are further classified according to the units related to the transmitter or receiver.

1.4
UNITS RELATED TO THE TRANSMITTER

Units related to the transmitter are as follows:

1.4.1 Radiant Emittance (W)

It is the total power radiated in all directions from unit area. It is equal to the radiant flux per unit area.

$$W = \frac{\text{Radiant flux}}{\text{Area}}$$

$$= \frac{\phi}{A}$$

Unit of radiant emittance is Wm^{-2}. The total radiant flux emitted from an element of surface area dA with a radiant emittance W is given by

$$\phi = \int W\, dA$$

The corresponding photometric quantity is luminous emittance. Its unit is lm m^{-2}.

1.4.2 Radiant Intensity (I)

Radiant intensity is the power radiated by a point source in a given direction into a unit solid angle (see Figure 1.2).

$$I = \frac{\phi}{\Omega}$$

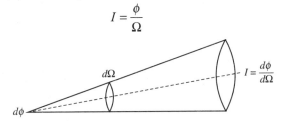

Figure 1.2 Power radiated by a point source per solid angle.

Unit of radiant intensity is watt per steradian (Wsr^{-1}). Corresponding quantity in photometry is luminous intensity. Its unit is lumen per steradian lm sr^{-1} or candela (cd)

$$\text{cd} = \frac{\text{lm}}{\text{sr}}$$

1.4.3 Radiance (L)

For a small plane source, the equivalent term to radiant intensity is radiance. Radiance is the power radiated from unit area into unit solid angle.

In Figure 1.3, if θ is the angle between the direction of radiation and normal to the surface, then the radiance

$$L = \frac{\phi}{\Omega A \cos\theta}$$

where the radiant flux ϕ leaving the surface is assumed to be uniform over all the surface and over all the solid angle.

Also, since $I = \phi/\Omega$, we can write

$$L = \frac{I}{A \cos\theta}$$

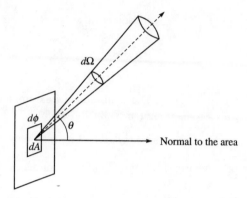

Figure 1.3 Power radiated from small plane into unit solid angle.

The radiance from an element of area dA over a small solid angle $d\Omega$ is

$$L = \frac{d^2\phi}{d\Omega\, dA \cos\theta}$$

or

$$L = \frac{dI}{dA \cos\theta}$$

Luminance is the corresponding quantity in photometry. Its unit is $cd\,m^{-2}$ (or $lm\,sr^{-1}\,m^{-2}$).

1.5
UNITS RELATED TO THE RECEIVER

1.5.1 Irradiance (*E*)

Irradiance is the total radiant power falling on unit area of the receiver. If the radiant power is equal on all over the surface, then

$$E = \frac{\phi}{A}$$

The unit of irradiance is Wm^{-2}. Photometric quantity corresponding to irradiance is illuminance. Unit of illuminance is Lux ($lm\,m^{-2}$)

$$lux = \frac{lm}{m^2}$$

1.5.2 Point Source

A small radiating surface can be regarded as a point source radiator, if the distance to the receiver is very large in comparison with the dimension of the radiating surface.

The power radiated by a point source with radiant intensity I in a given direction into a solid angle $d\Omega$ is

$$d\phi = I\,d\Omega$$

The total power radiated by a uniformly radiating point source is

$$\phi = \int_0^{4\pi} I\, d\Omega$$
$$= 4\pi I$$

Let us suppose an area dA is irradiated by a point source, which is located at a distance R from the source, and subtends a solid angle $d\Omega$ (see Figure 1.4). Let the normal to dA be inclined at an angle θ to the line that joins the source and the area dA. Then,

$$d\Omega = \frac{dA\cos\theta}{R^2}$$

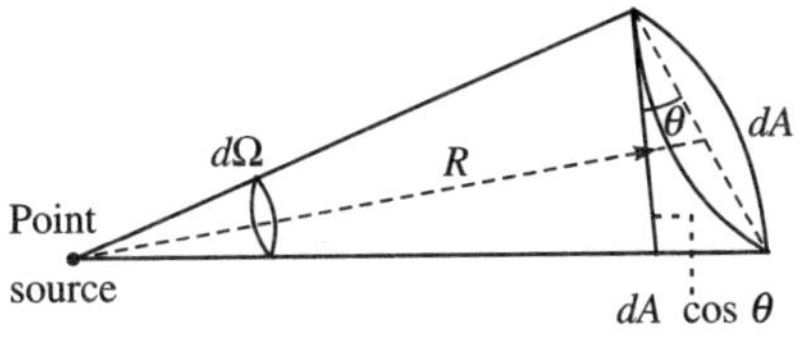

Figure 1.4 Inverse square law of illumination.

The power of radiation that falls on dA is

$$d\phi = \frac{I\, dA\cos\theta}{R^2}$$

By the definition of irradiance, the radiant power falling on the area of the receiver dA is

$$d\phi = E\, dA$$

i.e. $$I\, d\Omega = E\, dA$$

Or the irradiance brought about by a point source is

$$E = \frac{I\, d\Omega}{dA} = \frac{I\, dA\cos\theta}{dA\, R^2}$$
$$= \frac{I\cdot\cos\theta}{R^2}$$

Thus, the irradiance is inversely proportional to the square of the distance of the area from the source of radiation. This is known as the inverse square law of illumination.

1.6

LAMBERT'S LAW

The intensity of light emitted from a source is found to be different in different directions, and is given by

$$I_\theta = I_n \cos\theta \tag{1.1}$$

where I_θ is the intensity in a direction making an angle θ with normal to the surface and I_n is the intensity along the normal (Figure 1.5). Equation (1.1) shows that intensity in a particular direction is proportional to $\cos\theta$. The intensity is maximum along the normal and falls to zero in the direction tangent to the surface. This is called Lambert's Law or Lambert's cosine law of emission. A surface which obeys this law is called a uniformly diffusing surface (uniform diffuser or Lambert's radiator).

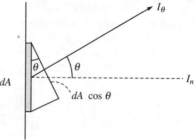

Figure 1.5 Lambert's cosine law.

A reflecting surface which obeys the cosine law for all directions of the incident light is called a uniformly diffused reflector. If such a surface reflects all the incident light it is said to a perfect diffuser. A thick layer of freshly prepared magnesium oxide on a plane surface forms an almost perfect diffuser.

1.7
EXTENDED SOURCE

In extended sources, it is assumed that the radiance is independent of angle. Such sources are known as Lambert's source and appear equally bright in all directions.

1.8
DIFFUSE REFLECTOR

Suppose the irradiance E falls on a small area ds. Then the total power of radiation falling on the surface is

$$d\phi_i = E\,ds$$

If the surface scatters or reflects a fraction k of this power, then the total power leaving the surface is given by

$$d\phi_s = k\,d\phi_i$$

It is important that the power leaves the surface in all directions. It is immeterial whether the surface radiates or scatters.

1.9
TOTAL RADIANT POWER FROM A SURFACE

Consider a small area ds of a uniformly diffusing surface of radiance L, so that the radiant intensity in the direction θ is given by

$$I = L\,ds\,\cos\theta$$

Draw a circle of radius r as shown in Figure 1.6, to calculate the radiant flux from a small strip of width $rd\theta$, between θ and $\theta + d\theta$.

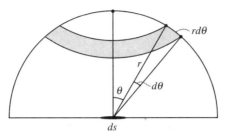

Figure 1.6 Radiant power from a surface.

The area of the strip

$$= 2\pi \text{ radius} \times \text{Width}$$
$$= 2\pi\, r \sin\theta \times r\, d\theta = 2\pi r^2 \sin\theta\, d\theta$$

Hence, solid angle subtended by this strip

$$d\Omega = \frac{\text{Area}}{(\text{distance})^2}$$
$$= 2\pi \sin\theta\, d\theta$$

Thus, the radiant flux $d\phi$ over the above solid angle becomes

$$d\phi = I\, d\Omega$$
$$= L\, ds \cos\theta \cdot 2\pi \sin\theta\, d\theta$$

∴ The flux radiated on a cone of angle θ is

$$\phi = \int_0^\theta 2\pi L\, ds \sin\theta \cos\theta\, d\theta$$
$$= \pi L\, ds \sin^2\theta$$

The total flux radiated from ds is obtained by putting $\theta = \pi/2$

$$\phi = \pi L\, ds$$

EXAMPLE 1.1 A plane surface is illuminated by a point source of radiant power ϕ kept at a distance r from the surface. Find the illumination on a point, at a distance d from the foot of the perpendicular from the source to the plane surface.

Solution By inverse square law

$$E = \frac{\phi \cos\theta}{SP^2}$$
$$= \frac{\phi \cos\theta}{r^2 + d^2}$$

But

$$\cos\theta = \frac{r}{\sqrt{r^2 + d^2}}$$

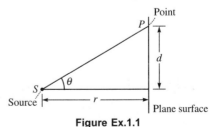

Figure Ex.1.1

or

$$\frac{\cos^2\theta}{r^2} = \frac{1}{r^2+d^2}$$

$$\therefore \quad E = \phi \frac{\cos^3\theta}{r^2}$$

EXAMPLE 1.2 Show that the radiant intensity I of a point source with radiant flux ϕ is given by

$$I = \frac{\phi}{4\pi}$$

Solutions We have,

$$I = \frac{d\phi}{d\Omega}$$

$$\phi = \int_0^{4\pi} I\, d\Omega$$

$$= I\, 4\pi$$

or

$$I = \frac{\phi}{4\pi}$$

EXAMPLE 1.3 A surface of area dS_2 is illuminated by a uniformly diffusing surface of area dS_1 whose illumination is L. Their normal makes an angle θ_1 and θ_2, respectively with respect to the line joining them and are separated by a distance r. Calculate the illuminance on dS_2.

Solution Luminant flux in the direction of θ_1 is

$$d\phi_1 = L \cdot dS_1 \cdot \cos\theta_1 \cdot d\Omega$$

where, $d\Omega$ is the solid angle subtended by dS_2 on dS_1 and is given by

$$d\Omega = \frac{dS_2 \cos\theta_2}{r^2}$$

Figure Ex.1.3

$$\therefore \quad d\phi_1 = \frac{L\, dS_1\, dS_2\, \cos\theta_1\, \cos\theta_2}{r^2}$$

$$= d\phi_2$$

Illuminance of ds_2

$$E = \frac{d\phi_2}{dS_2}$$

$$= L\frac{dS_1\, \cos\theta_1\, \cos\theta_2}{r^2}$$

EXAMPLE 1.4 A light source of luminous power 100 lm in suspended 6 m vertically above a horizontal surface. Calculate the illuminance at a point 8 m from the foot of the lamp.

Solution

$$E = \frac{\phi \cos \theta}{r^2}$$

$$= \frac{100 \times (6/10)}{6^2 + 8^2}$$

$$= 0.6 \text{ lm m}^{-2}$$

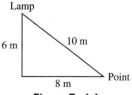

Figure Ex.1.4

REVIEW QUESTIONS

1. What is range of wavelength in the electromagnetic spectrum considered as optical radiation?
2. Distinguish between radiometry and photometry.
3. What is the responsivity of the human eye?
4. Write short notes on
   (i) Point source
   (ii) Extended source
   (iii) Diffuse reflector
5. Derive the inverse square law of light illumination.
6. Discuss the unit and the definition of various radiometric and photometric quantities.
7. What is a solid angle? How can it be calculated?
8. Calculate the total power radiated from a diffusing surface.

2
Elements of Optical Phenomena in Semiconductors

This chapter discusses various optical processes in semiconductors, the important property that is utilized in many optoelectronic devices like LEDs, Lasers, detectors etc. A complete description of these phenomena involves the theory of semiconductor physics and the theories of light-matter interaction, though we will consider optical process in semiconductor in a more simplified approach.

2.1
SEMICONDUCTOR PHYSICS

An intrinsic semiconductor such as silicon (germanium, etc.), which is in very pure form, contains negligibily by small amount of impurities. Each silicon atom shares its four valence electrons, with four neighbouring atoms forming four covalent bonds. In an n-type silicon, a substitutional pentavalent atom (e.g. phosphorous) with five valence electrons replaces a silicon atom and an electron is donated to the conduction band. This pentavalent atom is called a donor. When a trivalent atom such as boron with three valence electrons substitutes for a silicon atom an additional electron is accepted to form four covalent bonds around the boron and positively charged hole is created in the valence bond. This is p-type material and the impurity added is called acceptor.

In a semiconductor depending on temperature the carriers—electrons or holes—are distributed in the dopant energy levels in the respective bonds. The number of carriers at any energy level depend on the number of available states at that energy and is represented by density state function. The occupancy of states, either in the conduction or valence bands or in impurity level is determined by Pauli's exclusion principle. In equilibrium, the electron and hole occupation is represented by Fermi-Dirac (FD) statistics and the carrier densities can be evaluated from equilibrium Fermi level by using FD distribution function. Thus, the two important functions that determine the carrier distribution in a semiconductor are the energy distribution function and the density state function.

2.2
ENERGY DISTRIBUTION FUNCTION

The probability that an electron occupied in an electron state with energy E is given by FD distribution functions,

$$f_e(E) = \frac{1}{\left[\exp\dfrac{(E-E_F)}{kT}+1\right]}$$

where E_F is the Fermi energy.

The maximum value of the function is unity. Therefore, the probability of occupation of an energy level can never exceed unity or not more than one electron can occupy the same quantum state. Referring Figure 2.1, at $T = 0$ K, $f_e(E) = 1$, for $E < E_F$ and $f_e(E) = 0$ for $E > E_F$, so that all levels below the Fermi levels are occupied, while those above it are empty. At any temperature $f_e(E) = 1/2$ for $E = E_F$. Thus, the Fermi level can be defined as that energy level up to which all levels are occupied and above which all levels are empty at 0 K. At any temperature, $T > 0$ K. The probability of occupation at the Fermi level is 1/2. E_F is temperature dependent and its value decreases with the increase of temperature.

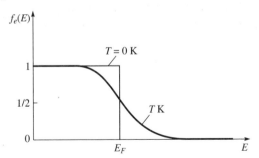

Figure 2.1 FD distribution function.

If the carriers have large energy due to heating or due to an applied field, i.e., if $(E - E_F) \gg kT$, the FD distribution function may be approximated by

$$f_e(E) = \exp\left[\frac{-(E-E_F)}{kT}\right]$$

$$= A\exp\left(\frac{-E}{kT}\right)$$

where,

$$A = \exp\left(\frac{E_F}{kT}\right)$$

Thus, the situation becomes similar to that of a Boltzmann distribution function and the restriction set by the Pauli's exclusion principle is relaxed. The hole distribution function $f_h(E)$ is given by

$$f_h(E) = 1 - f_e(E)$$

$$= 1 - \frac{1}{\exp\dfrac{(E-E_F)}{kT} + 1}$$

$$= \frac{1}{\exp\left(\dfrac{E_F - E}{kT}\right) + 1}$$

For $\quad E_F - E \gg kT$

$$f_h(E) \approx \exp\left\{\frac{-(E_F - E)}{kT}\right\}.$$

2.3
DENSITY OF STATES

The density of states function $N(E)\,dE$ gives the number of available quantum states in the energy interval between E and $E + dE$. For the conduction and valence bands the three-dimensional density states functions are

$$N_e(E) = \frac{4\pi}{h^3}(2m_e)^{3/2}(E - E_c)^{1/2}\,dE$$

and

$$N_h(E) = \frac{4\pi}{h^3}(2m_h)^{3/2}(E_v - E)^{1/2}\,dE$$

respectively, here E_c and E_v are conduction and valence band energies.

The density of electron and holes in the conduction and valence band can be obtained from

$$n_0 = \int_{E_c}^{\infty} N_e(E)\,f_e(E)\,dE$$

and

$$p_0 = \int_{-\infty}^{E_v} N_h(E)\,[1 - f_e(E)]\,dE$$

Thus, under Boltzmann's approximation, the density of electrons in the conduction band

$$n_0 = N_c \exp\left(\frac{E_F - E_c}{kT}\right)$$

and the density of holes in the valence band is given by

$$p_0 = N_v \exp\left(\frac{E_v - E_F}{kT}\right)$$

where
$$N_c = 2\left(\frac{2\pi m_e kT}{h^2}\right)^{3/2},$$

and
$$N_v = 2\left(\frac{2\pi m_h kT}{h^2}\right)^{3/2},$$

are the effective density of states in the conduction and valence band, respectively.

In intrinsic semiconductors, the electron concentration is equal to this hole concentration, since each electron in the conduction band leaves a hole in the valence band.

If we multiply electron and hole concentration we get,

$$n_i^2 = n_0 p_0 = N_c N_v \exp\left[\frac{-(E_c - E_v)}{kT}\right]$$
$$= N_c N_v \exp\left(\frac{-E_g}{kT}\right)$$

where, $E_g = E_c - E_v$ is the band gap energy. For intrinsic semiconductor, the carrier concentration is given by

$$n_i = 2\left(\frac{2\pi kT}{h^2}\right)^{3/2} (m_e m_h)^{3/4} \exp\left(\frac{-E_g}{kT}\right)$$

For $n_0 = p_0$, we obtain the Fermi level position measured from the valence band edge. If the intrinsic fermi level is denoted by E_{Fi}, we get

$$E_{Fi} = \frac{E_c + E_v}{2} + \frac{3}{4}kT \log\left(\frac{m_h}{m_e}\right)$$

Thus, the Fermi level of an intrinsic semiconductor material is close to the mid band gap (see Figure 2.2).

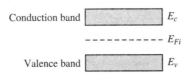

Figure 2.2 Valence band and conduction band.

2.4
EFFECT OF DOPING

At 0 K or very low temperatures the valence band is filled with electrons and the conduction band is empty. Under such condition there is no electrical conduction. For electrical conduction to occur covalent bonds has to be broken, for which the minimum energy required is band gap energy E_g. This energy can be provided by means of thermal, electrical, optical and a variety of other excitation techniques. In an intrinsic semiconductor without any dopant atoms the breaking of a covalent bond creates an electron-hole pair which moves in opposite direction under an electric field. Under thermal equilibrium the condition

$$n_0 = p_0 = n_i$$

is satisfied. Here, n_i is the intrinsic carrier concentration. Usually in semiconductors with $E_g \approx 1$ eV; n_i is too small for any practical use. Therefore, doping is used to increase the carrier concentration of n or p, and such semiconductors are called extrinsic semiconductors.

The position of the Fermi energy in a semiconductor is closely related to the doping-type and level of doping. In an n-type semiconductor the Fermi level is close to the conduction band while in a p-type semiconductor, Fermi level is close to the valence band.

In Figure 2.3, the energy required to excite on electron from the donor level into the conduction band is given by

$$E_D = E_c - E_d$$

where, E_d is the donor energy. Since the donor energy is very close to the conduction band energy, the concentration of electron is nearly equal to that of the electrons. Thus, even at moderately low temperatures, most of the electrons excited to the conduction band enhances the free electron concentration.

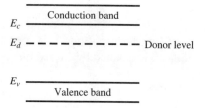

Figure 2.3 The energy level of the donor impurity atom.

In Figure 2.4, $E_A = E_a - E_v$ the energy required to create a hole. The acceptor level lies in the energy gap slightly above the valence band energy. The ionization process may represent a downward transition of a hole.

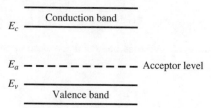

Figure 2.4 The energy level of the acceptor impurity atom.

2.5
QUASI-FERMI LEVELS UNDER CHARGE INJECTION

When excess charge carriers (electrons or holes) are injected by optical excitation or by applying electric field in a semiconductor, the system may not be in equilibrium and the concept of Fermi energy levels requires modification. Under external injection, the electron or hole occupancy can be described by a new non-equilibrium distribution function by the use of Quasi-Fermi levels under following assumptions:

(i) The electrons are essentially in thermal equilibrium in the conduction band and the holes are in equilibrium in the valence band. This means that the carriers are neither gaining nor losing energy from the crystal lattice.

(ii) The electron-hole recombination time is much larger than the time for the electrons and holes to reach equilibrium within the conduction band and the valence band, respectively.

The non-equilibrium distribution functions for electrons and holes can be represented by

$$f_n(E) = \frac{1}{1+\exp\left(\dfrac{E - E_{F_n}}{kT}\right)} \qquad (2.1)$$

and

$$f_h(E) = \frac{1}{1+\exp\left(\dfrac{E_{F_h} - E}{kT}\right)} \qquad (2.2)$$

respectively. Here, E_{F_n} and E_{F_h} are known as Quasi-Fermi levels for electrons and holes, respectively. When the excitation source is removed $E_{F_n} = E_{F_h} = E_F$. The deviation from the equilibrium is measured by $E_{F_n} - E_{F_h}$. If excess electrons or holes are injected into the semiconductor, the electron Quasi-Fermi level E_{F_n} moves towards the conduction band while the hole Quasi-Fermi level E_{F_h} moves towards the valence band. For non-degenerate case Eqs. (2.1) and (2.2) are:

$$f_n(E) = \exp\left(\dfrac{E_{F_n} - E}{kT}\right) \qquad (2.3)$$

and

$$f_h(E) = \exp\left(\dfrac{E - E_{F_h}}{kT}\right) \qquad (2.4)$$

The concentration of carriers can be determined from, density of state function as

$$n = \int_{E_c}^{\infty} N_e(E)\, f_n(E)\, dE$$

$$p = \int_{-\infty}^{E_v} N_h(E)\, f_h(E)\, dE$$

Using Eq. (2.3) and (2.4), we get, the non-equilibrium concentration as

$$n = N_c \exp\left(\dfrac{E_{Fn} - E_c}{kT}\right)$$

and

$$P = N_v \exp\left(\frac{E_v - E_{Fh}}{kT}\right)$$

2.6
BAND STRUCTURE

It is usual that the band structure of a semiconductor crystal is represented by a simple relation of the form $E = E(k)$, where E is the energy and k is the wave vector of the carrier. The relationship between E and k gives the band structure, which has the regions of allowed bands separated by a forbidden band gap. In a semiconductor at 0 K, the band that is filled with electrons is called the valence band, while the upper unfilled band is called the conduction band. The maximum of the valence band and the minimum of the conductor's band, for most semiconductors occurs at $k = 0$, i.e. the effective momentum is zero.

Figure 2.5 shows a simplified band structure of a typical semiconductor. In the figure, the valence band is concave downward and the conduction band is concave upward, which are separated by the energy gap E_g. The conduction band minimum and the valence band maximum occur at the zone centre $k = 0$. The zero energy level is chosen to lie at the top of the valence band (i.e., $E_v = 0$).

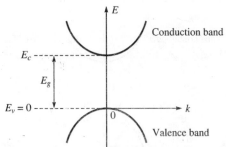

Figure 2.5 Band structure of a semiconductor.

The electron energy in the conduction band is related to the wave vector by

$$E_c(k) = \frac{\hbar^2 k^2}{2m_e^*} + E_g$$

where, m_e^* is the effective man of the electron and $\hbar = h/2\pi$. The energy of the valence band may be written as

$$E_v(k) = -\frac{\hbar^2 k^2}{2m_h^*}$$

where, m_h^* is the mass of the hole.

2.7
DIRECT BAND GAP SEMICONDUCTORS

In the band structure representation in some semiconductors the centre of the band occurs at $k = 0$, i.e. the valence band maximum and the zone centre ($k = 0$).

Such semiconductors are called direct band gap materials. In these materials, the upward or downward transition of electrons taken places without the involvement of a phonon, and there is no change in momentum. Consequently, in direct band gap materials, an electron raised to the conduction band by absorption will stay there for a short duration and recombine with a hole in the valence band and emit photons of energy equal to the band gap energy (see Figure 2.6).

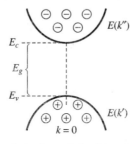

Figure 2.6 Band structure of direct band semiconductors.

$$\frac{hC}{\lambda} = E_g = E_c - E_v$$

They are optically active because the probability of recombination is very high so that the optical transitions are very strong. Some examples for direct bandgap semiconductors are GaAs, InP, InGaAs, etc.

The wave vector of the photon is given by

$$k_{ph} = \frac{2\pi}{\lambda}$$

2.8

INDIRECT BAND GAP SEMICONDUCTORS

In indirect bandgap semiconductors, the bottom of the conduction band does not occur at $k = 0$, but occur at some other k values while the maximum of the valence band occurs at a zone centre $k = 0$. Here, the upward or downward transition of carriers require a change in momentum $\Delta k = k'' - k'$, thus, there will be an involvement of phonons of energy E_p (see figure 2.7).

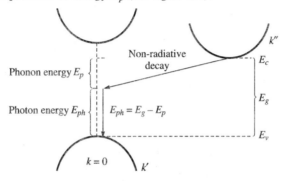

Figure 2.7 Transition in an indirect band gap semiconductor.

The electron in the conduction band at $k \neq 0$ cannot combine with a hole at $k = 0$, until a phonon with right energy and momentum is available. The hole will recombine non-radioactively and the excess energy is dissipated into the lattice as heat. In indirect band gap semiconductors, the non-radioactive processes reduces the probability of radioactive recombination, and hence, are not suitable materials for semiconductor light sources like LED or SDL, etc. In indirect band gap semiconductors optical transitions are very weak, and taken palce with the involvement of phonon to conserve the momentum.

2.9
ELECTRON HOLE PAIR CREATION AND RECOMBINATION

In a semiconductor, at absolute zero temperature, normally the valence band, is completely filled with electrons. At room temperature, the electron can be excited from the valence band to the conduction band as a result of their thermal energy. This creates an equal number of empty states in the valence band. These positively charged empty states are called holes. They are regarded as particles which behaves like electrons, but with a positive charge. With respect to the binding in the lattice, creation of a hole is equivalent to breaking a covalent bond. When electrons make such a transition, holes are left in the valence band with electrons in the conduction band, and e-h pairs are created. Pair formation essentially involves the rising energy of an electron from the valence band to the conduction band (see Figure 2.8). When excited by optical, electrical, thermal or by any other means—the electron in the conduction band and holes in the valence band drift in opposite direction and both thus contribute to the current known as electron current and hole current.

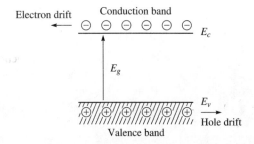

Figure 2.8 Electron-hole pair formation.

Electron-hole pairs can be created by irradiating light having suitable energy, in which a valence band electron raise to the conduction band. This process is called absorption. The process in which an electron in the conduction band makes a transition into a vacant space in the valence band giving up its excess energy is known as recombination. This is opposite to the absorption. Depending on the nature of recombination the excess energy is released as photons or phonons (qualized lattice vibrations) which are known as radiative and non-radiative recombination processes, respectively. The working of all optoelectronic devices are based on the creation and annihilation of electron-hole pair.

In a radiative recombination, a photon with energy equal to the band gap is released. The *e-h* pairs can also recombine without emitting light, instead they may emit either heat or a long wavelength photon together with a phonon. The recombination proces of emitting phonons is called non-radiative recombination. Non-radiative decay reduces the efficiency of optoelectronic devices.

In equilibrium, the electron-hole pair generation rate and the recombination rate are equal. At any temperature the probability of an electron recombining is proportional to the number of holes present. Thus, the recombination rate is proportional to the product of electron and hole concentration

$$R = Bnp$$

where, *B* is a constant of proportionality.

2.10
RECOMBINATION MECHANISMS

There are two types of recombination processes which are termed as direct transition (band-to-band transition) and indirect transition (defect centre transition). In the direct transition process, an electron in the conduction band makes a transition directly to the valence band and combine with a hole. In the indirect transition, recombination takes places via defect centre or traps in the forbidden energy gap which are associated with defect states caused by the presence of impurities or lattice imperfections (see Figure 2.9).

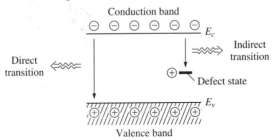

Figure 2.9 Recombination of electron and holes.

2.11
RADIATIVE RECOMBINATION EFFICIENCY

By optical excitation *e-h* pairs are generated, if the energy of the incident photon is sufficient to excite an electron from valence band to the conduction band, i.e., $h\nu \geq E_g$. Under steady state condition electron and holes are created and annihilated in pairs and an excess density $\Delta n = \Delta p$ is established. If the excess carriers are created in this way and if the exciting source is cut-off the carrier concentration will return gradually to their thermal equilibrium values due to the recombination of these carriers. The decay of carriers usually follow the relation

$$\Delta n(t) = \Delta n(0) \exp\left(\frac{-t}{\tau}\right)$$

where, Δn is the excess carrier density when the excitation source is switched off ($t = 0$) and τ is the lifetime of excess carriers, which is an important parameter on which the performance of an optoelectronic device depends on. Let τ_r and τ_{nr} be the lifetime for radiative and non-radiative decay, then

$$\frac{1}{\tau} = \frac{1}{\tau_r} + \frac{1}{\tau_{nr}}$$

Let R_r and R_{nr} are radiative and non-radiative recombination rate per unit volume, with the total recombination rate,

$$R_{sp} = R_r + R_{nr}$$

The radiative recombination efficiency, is defined as (also known as internal efficiency)

$$\eta_r = \frac{R_r}{R_{sp}} = \frac{R_r}{R_{nr} + R_r} = \frac{1}{1 + (R_{nr}/R_r)}$$

Thus, radiative recombination is maximum when non-radiate decay rate is small. Also, we can write

$$\tau_r = \frac{\Delta n}{R_r}$$

and

$$\tau_{nr} = \frac{\Delta n}{R_{nr}}$$

then,

$$\eta_r = \frac{\tau_{nr}}{\tau_r + \tau_{nr}}$$

$$= \frac{1}{1 + (\tau_r/\tau_{nr})}$$

To obtain high internal quantum efficiency τ_{nr} should be large enough, i.e. the non-radiative decay lifetime must be very large.

In the presence of excitation let the density of states be modified as

$$n = n_0 + \Delta n$$

and

$$p = p_0 + \Delta p$$

with $\Delta n \approx \Delta p$. For band to band transition spontaneous recombination rate is given by

$$R_{sp} = B_r np$$
$$= B_r(n_0 + \Delta n)(n_0 + \Delta p)$$
$$= B_r[n_0 p_0 + \Delta n(n_0 + p_0) + \Delta n^2]$$
$$= B_r n_0 p_0 + B_r \Delta n(n_0 + p_0 + \Delta n)$$

we can also express

$$R_{sp} = R_{sp}^0 + R_{sp}^{ex}$$

where, R_{sp}^0 and R_{sp}^{ex} are recombination rates without and with excitation. Thus

$$R_{sp}^{ex} = B_r \Delta n(n_0 + p_0 + \Delta n)$$

$\therefore$

$$\tau_r = \frac{\Delta n}{R_{sp}^{ex}} = \frac{1}{B_r(n_0 + p_0 + \Delta n)}$$

In the case of laser action $\Delta n \gg n_0, p_0$, then

$$\tau_r = \frac{1}{B_r \Delta n}$$

and for low level excitation $\Delta n \ll n_0, p_0$, then

$$\tau_r = \frac{1}{B_r(n_0 + p_0)}$$

In the case of intrinsic semiconductors $n_0 \approx p_0 \approx n_i$ and for low level injection

$$\tau_2 = \frac{1}{2 B_r n_i}$$

The value B_r depends on the property of the semiconductor and also whether the material has a direct or indirect band gap. Its value ranges from 10^{-11} to 10^{-9} cm^{3}s^{-1} and for direct band gap materials and 10^{-15} to 10^{-13} cm^{3}s^{-1} for indirect band gap materials.

2.12
ABSORPTION MECHANISMS IN SEMICONDUCTORS

In a semiconductor, the upward or downward transition of charge carriers between the energy band result into absorption or emission of electromagnetic energy which is the basis of all optoelectronic devices. Absorption is a process in which an electron in the lower state excited to a higher energy state by absorbing a suitable amount of energy. Some of the important types of absorption transition are discussed below:

2.12.1 Band to Band Transition

The simplest form of photon absorption is band to band absorption in which an electron in the valence band is excited to the conduction band (see Figure 2.10). Consider an electron rising from the top of the valence band to the bottom of the conduction band by the absorption of a photon.

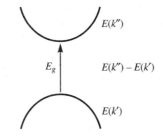

Figure 2.10 Band to band absorption.

Since there is no change in momentum,

$$k'' - k' = \Delta k = 0$$

where, k' and k'' are the wave vectors for the valence and the conduction band respectively.

22 Photonics: An Introduction

The energy of these bands are given by

$$E(k'') = E_g + \frac{\hbar^2 k''^2}{2m_e^*}$$

and

$$E(k') = -\frac{\hbar^2 k'^2}{2m_h^*}$$

Since $k'' = k' = 0$, the energy of absorbed photon is given by

$$E_{ph} = E_g = E(k'') - E(k')$$

2.12.2 Indirect Transition

In indirect transition, there is a change in the momentum involved, so that the transition can occur only by the emission or absorption of a phonon when the photon energy is absorbed. The conservation of momentum is given by

$$k' + k_{ph} = k'' \pm k_p$$

where k_{ph} and k_p are the wave vectors of the photon and phonon, respectively.
Since, $k_{ph} \ll k_p$,

$$k'' - k' \approx \pm k_p$$

In Figure 2.11, the energy of the photon absorbed is given by

$$\hbar\omega = E_c - E_v \pm E_p$$

where, E_p the phonon energy, + sign corresponds to phonon emission while − sign corresponds to phonon absorption.

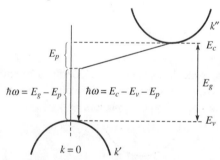

Figure 2.11 Indirect transition.

2.12.3 Low Energy Absorption

Several types of transitions take place at very small energies which give rise to resonances in the long wavelength region of the absorption spectra. Some low energy absorption processes are

   (i) Impurity band absorption
  (ii) Intraband absorption
 (iii) Free carrier absorption

Impurity Band Absorption

At very low temperature, impurity levels are usually filled with their respective carriers. These carries can be excited to the band edges as shown in Figure 2.12 by absorbing photons of appropriate energy. These transitions are known as low energy donor band and acceptor band absorptions. The energy of absorption corresponds to the infrared region of the optical spectrum.

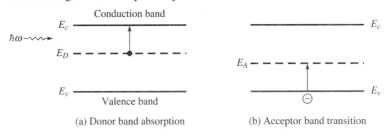

(a) Donor band absorption   (b) Acceptor band transition

Figure 2.12

Intraband Transitions

In most semiconductors, due to spin-orbit interactions at the zone centre, the valence band structure consists of sub-bands named as light hole (LH) band, heavy hole (H, H) band and split off (SO) band. In a *p*-type material absorption of a photon with right energy can result in transition from LH to HH, SO to HH and SO to LH bands depending on the doping and temperature. These are known as intraband absorption transitions. These absorption are not observed in *n*-type semiconductors.

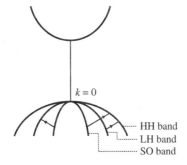

Figure 2.13   Intraband absorption

Free Carrier Absorption

In free carrier absorption a free electron within a band absorbs a photon and raised to a higher energy. The momentum is conserved and the momentum change is provided by photon or phonon or impurity scattering. Free carrier absorption usually manifests in the long wavelength of the spectrum.

Donor-Acceptor Absorption

As shown in Figure 2.14, in a semiconductor, both donor and acceptor state may be simultaneously present. It is possible to raise an electron from the acceptor to the donor level by absorbing a photon.

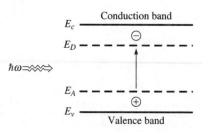

Figure 2.14 Donor-acceptors absorption.

The photon energy required is given by

$$\hbar\omega = E_g - E_D - E_A + \frac{q^2}{\varepsilon_0 \varepsilon_r R}$$

where, R is the separation between the impurity centres. The $q^2/\varepsilon_0\varepsilon_r R$ represents the coulomb interactions between the donor and acceptor in substitutional sites.

Impurity Level-band Absorption

Transition can take place between the impurity level and the opposite band edge as in Figure 2.15.

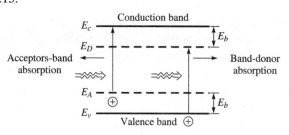

Figure 2.15 Impurity-band absorption.

The photon energy absorbed is given by

$$\hbar\omega = E_g - E_b$$

2.13
EXCITONS

In pure semiconductors, electron and holes produced by the absorption of a photon can be bound together because of the coulombic interaction of electron and hole. Such a bound electron-hole pair is called an exciton. It can be picture as an electron and a hole rotating about their common centre of mass. The binding energy of the exciton is

$$E_{ex}^l = -\frac{13.6}{l^2} \frac{m_r^*}{m} \left(\frac{1}{l_r}\right)^2 \text{ (eV)}$$

where, m_r^* is the reduced effective mass of the exciton given by

$$\frac{1}{m_r^*} = \frac{1}{m_e^*} + \frac{1}{m_h^*}$$

l is an integer ε_r is the relative permittivity and m is the rest mass of electron.

As shown in Figure 2.16, in direct band gap semiconductors the total energy of the exciton is given by

$$E_{ex} = \frac{\hbar^2 k_{ex}^2}{2(m_e^* + m_h^*)} - E_{ex}^l$$

$\hbar^2 k_{ex}^2 / 2(m_e^* + m_h^*)$ represents the KE of the exciton.

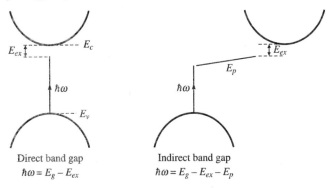

Figure 2.16 Exciton absorption.

Thus, for exciton excitation, the photon energy.

$$\hbar\omega = E_g - E_{ex}$$

In indirect band gap semiconductors exciton absorption takes place accompanying a phonon absorption or emission at energy.

$$\hbar\omega = E_g - E_{ex} \pm E_p$$

The excitons can move through a crystalline lattice. Hence, they provide an important means of transferring energy from one point to another in the material.

2.14
EMISSION IN SEMICONDUCTORS

When a photon is absorbed in a semiconductor an electron is raised from a lower level to a higher energy level. The energy of the absorbed photon being the difference between the energy of two levels involved. The excited electron can then make a downward transition to the lower energy state, i.e. the electron recombine with a hole emitting electromagnetic radiation. The energy of the radiation emitted is equal to the energy difference between the higher and lower energy states. The emission process in semiconductors can take in many ways, some of them are explained as follows:

2.14.1 Direct Band Transition

In direct band gap semiconductor, an electron in the conduction band recombine with a hole in the valence band (see Figure 2.17). The energy of the photon is

$$\hbar\omega = E_g$$

the band gap energy and $\Delta k = k'' - k' = 0$, i.e., there is no momentum change.

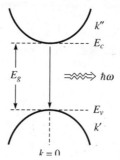

Figure 2.17 Direct band emission.

2.14.2 Indirect Band Transition

In this case as $k'' - k' = \Delta k \neq 0$, to conserve momentum in addition to the photon emission, interaction of phonon is also involved.

The photon energy is given by

$$\hbar\omega = E_g \pm E_p$$

where E_p is the phonon energy (+) for phonon annihilation and (−) for creation. Usually $E_p \approx 0.01$ eV.

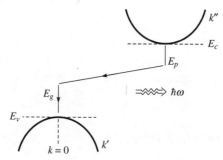

Figure 2.18 Indirect band emission.

2.14.3 Impurity Band Emission

These transitions are associated with the impurity states in the semiconductor. In indirect band gap semiconductor, emission via the impurity state can be utilized to increase radiative efficiency. Different types of impurity band emission transitions are as follows:

Donor-Acceptor Transition

Here an electron in the donor state recombine with a hole in the acceptor state [Figure 2.19(a)]. By emitting energy

$$\hbar\omega = E_D - E_A$$

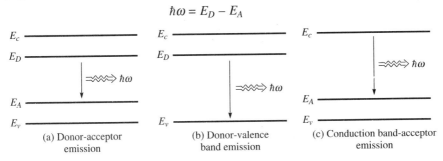

(a) Donor-acceptor emission
(b) Donor-valence band emission
(c) Conduction band-acceptor emission

Figure 2.19 Impurity state emission.

Donor-Valence Band Transition

An electron in the donor state recombine with a hole in the valence band [see Figure 2.19(b)]. The energy of the emitted photon is

$$\hbar\omega = E_D - E_v$$

Band Acceptor Emission

An electron in the conduction band recombine with an acceptor state hole emitting radiation of energy.

$$\hbar\omega = E_c - E_A$$

Exciton Emission

The band edge electron and hole are bound together to form an *e-h* pair called exciton by absorption of a photon. The recombination of *e-h* will produces a sharp emission lines corresponding to exciton energy. For direct band semiconductors, the energy of the emitted photon is given by

$$\hbar\omega = E_g - E_{ex}$$

as shown in Figure 2.20.

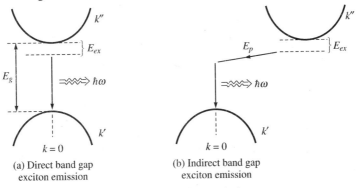

(a) Direct band gap exciton emission
(b) Indirect band gap exciton emission

Figure 2.20 Exciton emission.

In indirect bandgap semiconductors to conserve momentum phonon is also involved with energy E_p.

$$\hbar\omega = E_g - E_{ex} \pm E_p$$

Deep Level Emission

Lattice defects such as interstitials, vacancies or substitutional impurities, can act as carrier recombination centres or trapping centres and can give rise to deep level transitions. The excess energy of carriers recombining at these levels is carried out by phonons. Though radiative emission takes place, such transition degrade the radiation efficiency, and adversely effect device performance.

2.15
FRANZ-KELDYSH EFFECT

A strong electric field above 10^5 V/cm when applied to a semiconductor, the valence band electrons can easily tunnel to the conduction band. The result is that in the presence of such a strong electric field, the absorption of a photon takes place with energies less than the band gap energy $\hbar\omega < E_g$. This decrease in the absorption energy in the presence of an electric field is known as Franz-Keldysh effect.

2.16
STARK EFFECT

The change in the energy of atomic levels in the presence of an electric field is called Stark shift. As a result, there is a splitting in the energy levels. In linear Stark effect, the energy shift is directly proportional to the electric field. The dependence of the energy shift on higher orders of shift such as second order or cubic order etc., are known as non-linear Stark effect.

2.17
STOKES SHIFT IN SEMICONDUCTOR

When a photon is absorbed by a semiconductor, it is normally expected that the emission is also occuring at the same wavelength. However, due to the presence of defects, impurities, deep levels, etc., it is found that the emission spectrum shifted towards the longer wavelength as compared with the absorption. This shift in the transition energy between absorption and emission is known as the Stokes shift.

To explain the Stokes shift in semiconductors consider the Figure 2.21, which represents the ground state and excited state of an electron of an impurity atom as a function of nearest neighbour separation. Note that the minimum of the two curves occur at different nearest neighbour separation. Also the radiative processes, absorption and emission, take place very rapidly.

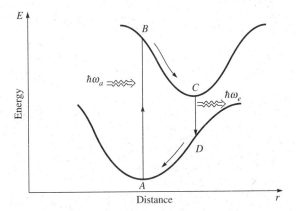

Figure 2.21 Stokes shift in semiconductors.

When the atom is at the equilibrium position A the absorption of a photon of energy $\hbar\omega_a$ excite an electron to the upper state B very rapidly with no change in momentum ($\Delta k = 0$). As B is not an equilibrium of the upper level, the electron relaxes to C, the minimum energy position of the state by phonon emission (lattice vibration). The electron then radiatively recombine with a hole emitting photon of energy $\hbar\omega_e$ and return to D a non-equilibrium position in the lower state. It then decays to the minimum energy of this level by phonon emission. The energy difference between the absorbed and emitted photon $\hbar\omega_a - \hbar\omega_e$ is known as the Stokes shift (or Franc-Condon shift).

The Stoke shift is utilized in fluorescent lamps where UV radiations from the electric discharge is allowed to absorb the fluorescent material coating given inside the lamp, which then converted into visible radiation thereby increases luminous efficiency.

2.18
AUGER RECOMBINATION

In a semiconductor, the recombination rate depends on the carrier concentration and can be expressed as

$$R(n) = An + Bn^2 + Cn^3$$

Here the first terms represent the non-radiative recombination associated with the defect centres and is called SRH (Shockley, Read, Hall) recombination which depends on the density of charge carriers. The second term contribution is due to spontaneous radiative recombination (proportional to np) and the last term is the Auger recombination rate. Auger recombination is a non-radiative recombination process involving three charge carriers. Depending on the final carrier—whether is an electron or a hole, the Auger recombination rate is proportional to np^2 or pn^2, respectively. In the process, the excess energy is released by recombination of an e-h pair is transferred to a third carrier as kinetic energy which is then raised in deep energy level in the respective band, and finally thermalizes back to the bottom of the band.

Different Auger recombination mechanism are shown in the Figure 2.22. For example, *e-h* pairs recombine emitting photons and the excess energy released is transformed as kinetic energy to the subbands of impurity states which is then excited.

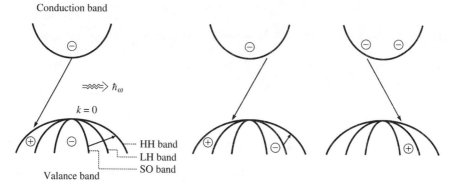

Figure 2.22 Auger recombination
HH—Heavy hole, LH—Light hole, SO—Split off (Spin-interaction states in a band)

The rate of Auger process increases with the temperature and also larger in small band gaps. It also depends on concentration of excess charge carriers. Auger processes are dominant in heavily doped narrow band gap semiconductors. This process limits the performance of semiconductor lasers which are made of such materials.

EXAMPLE 2.1 In a semiconductor, the band gap is 1 eV. Assuming band to band transition calculate the wavelength and the wave vector of the photons emitted.

Solution
$$\frac{hc}{\lambda} = E_g$$

$$\therefore \quad \lambda = \frac{6.6 \times 10^{-34} \times 3 \times 10^8}{1.6 \times 10^{-19}} = 1.237 \times 10^{-6} \text{ m}$$

$$k = \frac{2\pi}{\lambda} = 5.1 \times 10^6 \text{ m}^{-1}$$

EXAMPLE 2.2 In a semiconductor with a carrier density $n \approx p \approx 10^{17}$ cm^{-3}, under low level injector the *e-h* radiative lifetime is 2.5×10^{-7} sec. Calculate the recombination rate.

Solution
$$\tau = \frac{1}{B_r(n+p)}$$

$$B_r = \frac{1}{2.5 \times 10^{-7} \times 2 \times 10^{17}} = 2 \times 10^{-11} \text{ cm}^3 \text{ s}^{-1}$$

EXAMPLE 2.3 The effective mass of electron and hole in silicon is 0.26 m and 0.33 m, respectively where $m = 9.1 \times 10^{-31}$ kg, the mass of the electron. Calculate the effective reduced mass of the electron-hole system.

Solution
$$\frac{1}{m_r^*} = \frac{1}{m_e^*} + \frac{1}{m_h^*}$$

∴
$$m_r^* = \frac{m_e^* \times m_h^*}{m_e^* + m_h^*} = 0.145 \, m$$
$$= 1.32 \times 10^{-31} \, \text{kg}$$

EXAMPLE 2.4 An exciton of reduced mass 0.15 m is form in a semiconductor of relative permittivity 11.56. Calculate the binding energy of the exciton (where $m = 9.1 \times 10^{-31}$ kg, the mass of electron).

Solution
$$E_{ex} = -13.6 \frac{m_r^*}{m} \left(\frac{1}{\varepsilon_r}\right)^2 \, (\text{eV})$$
$$= -0.015 \, \text{eV}$$

EXAMPLE 2.5 Calculate the wavelength of the light emitted if the band gap is 2.24 eV.

Solution
$$\frac{hc}{\lambda} = E_g$$

∴
$$\lambda = \frac{hc}{E_g} = 552.4 \, \text{nm}$$

EXAMPLE 2.6 The radiative and non-radiative recombination lifetimes of the minority carries in the active region of an LED are 80 ns and 120 ns, respectively. Calculate the total carrier recombination lifetime and the internal quantum efficiency.

Solution
$$\frac{1}{\tau} = \frac{1}{\tau_r} + \frac{1}{\tau_{nr}}$$
$$\tau = \frac{\tau_r \cdot \tau_{nr}}{\tau_r + \tau_{nr}}$$
$$= 48 \, \text{ns}$$
$$\eta_r = \frac{1}{1 + (\tau_r/\tau_{nr})} = 0.6$$

EXAMPLE 2.7 Work function of cesium is 1.9 eV. Calculate the cut of wavelength.

Solution
$$\lambda = \frac{\hbar c}{\phi} = 651.3 \, \text{nm}$$

EXAMPLE 2.8 Calculate the radiative recombination life time in GaAs having injection level of $p_0 \approx n_0 = 10^{18}$ cm^{-3} and $B_r = 7 \times 10^{-10}$ cm^{3}s^{-1}.

Solution
$$\tau_r = \frac{1}{B_r(n_0 + p_0)}$$
$$= 7.14 \times 10^{-10} \, \text{s}$$

REVIEW QUESTIONS

1. Explain the concept of quasi Fermi level in a semiconductor under irradiation.
2. Write down quasi Fermi distribution functions for electrons and holes in a semiconductor.
3. Derive the expression for charge carriers in a semiconductor from the concept of quasi Fermi energy level.
4. Derive the expression for excess charge carrier concentration of a semiconductor under excitation.
5. Define minority carrier recombination life time in a p-type/n-type material.
6. What is band structure?
7. Explain direct and indirect band gap semiconductors in terms of band structure. Give some examples.
8. Compare the properties of direct and indirect band gap semiconductors.
9. Explain what do you mean by radiative and non-radiative recombination.
10. Why Si and Ge are not suitable for the realization of light sources in optoelectronics.
11. Which of the semiconductor type-direct or indirect band gap—is more suitable for making LED and SDL and Why?
12. Explain with energy level diagram of the following:
    band to band absorption
    indirect band absorption
    low energy absorption
    donor acceptor absorption
    impurity band absorption
    exciton absorption
13. What is an exciton?
14. Explain electron-hole recombinations using energy level diagram.
15. Define inernal quantum efficiency.
16. Define radiative recombination efficiency.
17. What is a hole current and electron current?
18. Mention any three types of excitation mechanism that can be used to produce e-h pairs in a semiconductor.
19. What do you mean by absorption?
20. What is recombination?
21. What are phonons?
22. Explain direct and indirect band recombination.
23. Explain the recombination mechanism in direct and indirect band gap semiconductors using energy level diagram.
24. What is the condition required to obtain high internal quantum efficiency?
25. Derive the expression for recombination lifetime for band to band transition.
26. Explain the absorption mechanism in semiconductors
27. Explain the emission transition in semiconductors.
28. Explain Franz-Keldysh effect.
29. Explain Stark effect in semiconductors.
30. Explain Stokes shift in semiconductors.
31. What is Auger recombination?
32. How Auger effect affects the efficiency of optoelectronic light sources?
33. What is SRH effect?

3

Light Emitting Diode

3.1
INTRODUCTION

Light Emitting Diode (LED) is a *p-n* junction device that converts electrical energy into optical radiation by a process known as injection luminescence (IL) or electro luminescence (EL). It is operated under forward biased condition which injects a large number of electrons into the empty conduction band. When these injected electrons in the conduction band radiatively recombine with holes in the valence band, light is emitted by spontaneous emission. Light emitting diode has wide applications as a light source in display devices, optical communications, sensors, indicator lamps, etc. First LED fabricated with silicon carbide was demonstrated in 1904. Recently, many LEDs operating in the ultraviolet to far infrared wavelength region using different semiconductor materials have been developed.

The most important advantages of the LED as a light source are simple fabrication procedure, simple drive circuitry and low cost. They have linear current-output light power characteristics and also are operated at low input powers. As compared with semiconductor diode laser, the output characteristics are less sensitive to temperature.

3.2
LUMINESCENCE

Luminescence is a radiative electron-hole recombination process in solid when some form of energy is applied to it. The excess energy associated with the recombination is imparted as photon with energy equal to the band gap, i.e.

$$h\nu = E_g$$

or

$$\frac{hc}{\lambda} = E_2 - E_1$$

In order to produce luminescence, the electrons from the lower state has to be excited to the upper level by supplying energy such as by optical excitation, electrical, electron beam bombardment etc. According to the method of excitation by which the electron-hole pairs are created, there are various types of luminescences as follows:

3.2.1 Photoluminescence

Photoluminescence is the luminescence process involving radiative electron-hole recombination that arises from the absorption of photon.

3.2.2 Cathodoluminescence

Cathodoluminescence is the luminescence process involving radiative electron-hole recombination as a result of excitation by electron beam bombardment.

3.2.3 Electroluminescence

Electroluminescence is the luminescence produced by radiative recombination processes resulting from the application of ac or dc electric field. Luminescence may persist for a time even the excitation source is switched off. If it exists for a duration equal to the lifetime of the transition between the upper and lower energy level, then it is known as fluorescence. If the luminescence exists for a longer time even after the excitation source is removed, the phenomena is called phosphorescence. The material exhibiting phosphorescence are called phosphors.

3.3
ELECTROLUMINESCENCE IN p-n JUNCTION

Electroluminescence from a p-n junction injecting minority charge carriers by applying a forward bias to the junction is known as injection luminescence. Under forward bias a large number of minority charge carriers are injected into the opposite sides of the junction. This is known as minority carrier injection. This creates a large number of electrons in the conduction band, which are then radiatively recombine with holes in the valence band, resulting into spontaneous emission. In the case of direct band gap semiconductors, recombination happens radiatively, so that the wavelength is given by

$$\frac{hc}{\lambda} = E_g$$

In the case of indirect band gap material, there will be non-radiative decay paths also, in which the excess energy of the carriers dissipated as heat in the lattice.

The process of light emission in a p-n junction can be explained as follows: In the p-n junction, the holes and electrons are the majority charge carriers in p-type and n-type material, respectively. When a forward bias is applied these majority carries crosses the depletion region and increases the minority carrier population in the opposite region (see Figure 3.1). These minority carriers diffuse away from the junction and recombine with majority carriers and emit light. Enhancing the local minority carrier population larger than the normal under forward bias is termed as minority carrier injection.

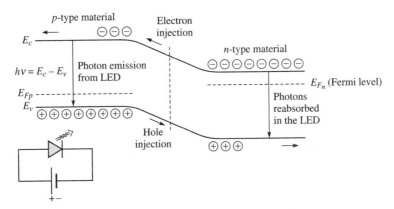

Figure 3.1 Minority carrier injection in a p-n junction under forward bias.

The electron-hole recombination rate in a forward biased *p-n* junction leading emission in LED under several conditions are listed as follows:

Strong Injection

When a high density of both electrons and holes are injected, the spontaneous recombination rate

$$R_{sp} = \frac{n}{\tau_0} \approx \frac{p}{\tau_0}$$

where τ_0 is the radiative lifetime.

Minority Carrier Injection

When the material is heavily doped with $n \gg p$ (the holes are injected into the heavily doped *p*-region). The *e-h* recombination rate is given by

$$R_{sp} = \frac{1}{\tau_0}\left(\frac{m_r^*}{m_e^*}\right)^{3/2} p$$

For $p \gg n$, the electrons are injected into the heavily doped *p*-region, then

$$R_{sp} = \frac{1}{\tau_0}\left(\frac{m_r^*}{m_h^*}\right)^{3/2} p$$

The recombination rate is proportional to the minority carrier density.

Weak Injection

At low level injection, the probability to recombine an electron with a hole is low. The recombination rate is given by

$$R_{sp} = \frac{1}{2\tau_0}\left(\frac{2\pi\hbar^2 m_r^*}{kT m_e^* m_h^*}\right)^{3/2} (\Delta n p_0 + \Delta p n_0)$$

The total recombination lifetime τ_0 is given by

$$\frac{1}{\tau_0} = \frac{1}{\tau_r} + \frac{1}{\tau_{nr}}$$

The internal quantum efficiency is given by the fraction of radiative recombination rate to the total recombination rate and is given by

$$\eta_{qe} = \frac{R_r}{R_{tot}} = \frac{\tau_{nr}}{\tau_{nr} + \tau_r} = \frac{1}{1 + (\tau_r/\tau_{nr})}$$

To achieve high internal quantum efficiency, the non-radiative lifetime τ_{nr} must be large. In a high quality direct band gap semiconductors, η_{qe} is close to unity, whereas in indirect band gap semiconductors it is of the order of 10^{-3} to 10^{-2}. The efficiency decreases with the increase in temperature.

3.4
RADIATIVE RECOMBINATION IN LED

Radiative recombination in LEDs occur predominantly by following mechanisms.

Interband Transitions

This corresponds to direct and indirect band transitions in which both, the energy of the photons emitted is given by

$$\hbar\omega = E_c - E_v = E_g$$
$$\hbar\omega = E_g \pm E_p$$

For direct and indirect transitions, respectively. In the second case phonon interaction is also involved.

Recombination via Impurity Centres

Three types of recombinations involving impurity levels are
    (a) conduction band-acceptor transitions
    (b) donor level-valence band transitions
    (c) donor-acceptor level transitions

Exciton transitions

Recombination of e-h pairs via excitons of energy E_{ex} will emit photons of energy

$$\hbar\omega = E_g - E_{ex}$$

and
$$\hbar\omega = E_g - E_{ex} \pm E_p$$

in direct and indirect band gap materials respectively.

3.5
LED DRIVE CIRCUIT

In a light emitting diode electrons and holes are injected as minority carriers across the junction under forward bias and they recombine either by radiative or non-radiative recombination. The diode is designed such that the radiative recombination is made strong as possible.

In dc biasing (Figure 3.2) a source voltage V_b is applied across the LED through a resistance R_S, which limits the current I_L. The value of R_S can be obtained by

$$V_b = I_L R_S + V_L$$

or

$$R_S = \frac{V_b - V_L}{I_L}$$

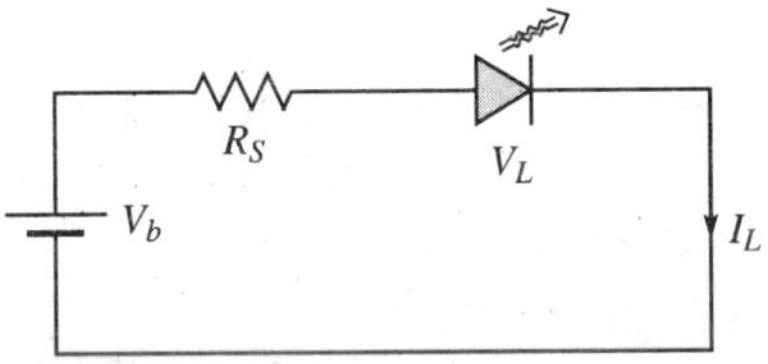

Figure 3.2 LED dc drive circuit.

The forward bias voltage depends on the band gap of the semiconductor material. The operating voltage vary form 1 V to 2 V and the current typically 10 mA to 100 mA.

3.6
LED CHARACTERISTICS

Performance of a light emitting diode is specified by following characteristics.
  (i) Spectral response characteristics (wavelength-intensity relationship)
  (ii) Current-light power characteristics
  (iii) Current-voltage characteristics
  (iv) Time-light power characteristics
  (v) Temperature dependence of light power
  (vi) Temporal response

Usually these will be provided by the manufacturer.

3.6.1 Spectral Response

Spectral response is a plot of relative intensity of light output of the LED as a function of wavelength. It is a measure of wavelength range over which the LED is emitting light.

The spread in wavelength of the output is specified as full width at half maximum (FWHM). FWHM is the width of the spectral response curve where the relative intensity falls to half of the maximum intensity.

The wavelength corresponding to the maximum relative intensity, the spectral response and FWHM depend on the band gap energy of the semiconductor and the doping level. They also depend on the working temperature. For AlGaAs, FWHM is about 40 nm, with maximum relative intensity at about 830 nm.

38 Photonics: An Introduction

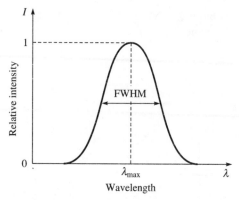

Figure 3.3 Spectral response of a LED.

3.6.2 Current-Light Power Characteristics

This is one of the most important characteristics of a LED. The current-light power (I-P_{ph}) characteristics curves are almost linear at low operating conditions. At high currents depending on the material properties non-linearities are also observed. This is because the light output starts to saturate as the LED heated up, so that the radiative recombination efficiency decreases.

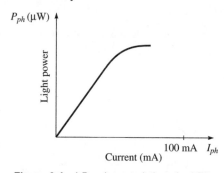

Figure 3.4 I-P_{ph} characteristics of a LED.

When the LED is forward biased, a fraction of the charge carriers is converted into light. The power of light that emitted from the diode is

$$P_{ph} = \frac{n}{t} h\nu$$

where n/t is the number of photons emitted in unit time.

But the efficiency

$$\eta = \frac{\text{Photons/Time}}{\text{Charge carriers/Time}}$$

$$= \frac{n/t}{N/t} = \frac{n/t}{I/e}$$

(Since total charge $Q = It$, $\therefore$ $eN = It$, or $N/t = I/e$)

$$\therefore \quad \frac{n}{t} = \eta \frac{I}{e}$$

Hence,
$$P_{ph} = \eta \frac{I}{e} h\nu$$

At very large currents stimulated emission may start where the photon density will become large. Such LEDs are called super luminescent light emitting diodes, and behaves similar to laser diodes.

3.6.3 I-V Characteristics

The I-V characteristics of a LED can be expressed as

$$I = I_0 \left[\exp\left(\frac{eV}{\beta kT}\right) - 1 \right]$$

where, I_0 is the reverse saturation current. It describes the electrical operating condition of the device and is as shown in Figure 3.5.

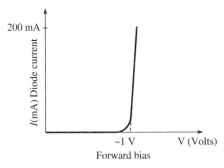

Figure 3.5 I-V characteristics of a LED.

3.6.4 Output Power-Time Characteristics

It specifies the output light power of a LED as a function of operating hours, which gives lifetime of the device. Usually the light power reduces with prolonged use. For AlGaAs LED, the loss in light power is about a few per cent after operation of about 8000 hours (Figure 3.6).

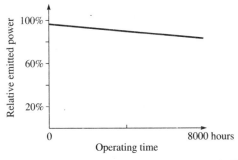

Figure 3.6 Output power-time characteristics of LED.

3.6.5 Temperature Dependence on Light Emission

The band gap energy of a semiconductor decreases with the increase in temperature. As a result peak of the emission spectrum of the LED will shift to higher wavelength region as temperature increases. This shift is approximately 0.35 nm/K for GaAs and 0.6 nm/K for InGaAsP LEDs. The FWHM of the emission spectrum also increases (i.e. spreads more) with the temperature.

As the temperature increases, a large number of charge carriers can leak across the active region. This leakage cause a large leakage current which will reduce the optical power. Auger processes, which are strongly temperature depended non-radiative process also reduce the optical power of the LED. Considering these two effects—carriers leakage and auger process—the optical output can be expressed as

$$I_{ph}(T) = I_{ph}(0) \exp\left(\frac{-T}{T_1}\right)$$

where T_1 is a temperature that depends on the band gap, and LED design should be so as to ensure temperature independence.

3.6.6 Temporal Response

An important application of LED is in communication, where the response time—how fast the device converts the electrical signal into an optical signal—is an important factor. A response time of the order of nanosecond is desirable for the communication purpose. The response time depends on the carrier recombination time, and is determined by the following factors:

Junction Capacitance

This can be reduced by decreasing the diode area and by chosing proper doping level.

Charge Storage and Diffusion Capacitance

This is due to the charge storage and the diffusion capacitance under forward biasing. This can be reduced by modulating at high speed.

The temporal response function of LED can be expressed as

$$r(\omega) = \frac{1}{(1 + \omega^2 \tau^2)^{1/2}}$$

where ω is the modulation frequency and τ is the minority carrier life time. The temporal response function represents the variation of optical power with the modulation frequency, and is as shown in Figure 3.7.

The frequency range over which the optical power reduce to half is called optical bandwidth and that over which the optical power reduced to 0.707 is called the electrical band width. Above equation shows that for high frequency response the recombinations lifetime τ should be made as small as possible.

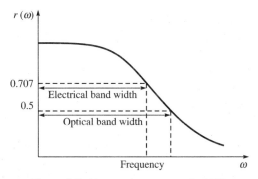

Figure 3.7 Temporal response of a LED.

3.7

LIGHT OUTPUT FROM LED AND OPTICAL RESPONSIVITY

When a forward bias is applied to the LED light is emitted from it as a result of recombination of electrons and holes. In direct band gap materials the energy of the emitted photon is equal to the band gap energy between the conduction and the valence band.

$$h\nu = E_g = E_c - E_v$$

The number of photons emitted per second depend on the probability of electron-hole recombination. The optical power is

$$P = \frac{n}{t} h\nu$$

The quantum efficiency

$$\eta = \frac{\text{Number of photons emitted per second}}{\text{Number of charge carriers produced per second}}$$

$$= \frac{n/t}{N/t}$$

The optical responsivity

$$R = \frac{\text{Optical power}}{\text{Injection current}}$$

$$R = \frac{P}{I}$$

$$= \frac{\frac{n}{t} h\nu}{\frac{Ne}{t}}$$

or

$$R = \eta \frac{h\nu}{e}$$

The external quantum efficiency, η_o is given by

$$\eta_o = \frac{\text{Total radiant output power}}{\text{Total input power}}$$

The total radiant power

$$W = h\nu \times (\text{photon emission rate})$$

3.8
EFFICIENCY OF LED

Luminescence efficiency of a light emitting diode depends on three different processes.

(i) Injection (or excitation) processes
(ii) Recombination processes
(iii) Extraction processes

Each of these processes become very efficient under suitable conditions, and we can characterize the efficiency of these processes as injection efficiency (η_i), recombination efficiency (η_r) and extraction efficiency (η_e) respectively. The overall efficiency is called the external efficiency, given by the product as

$$\eta_0 = \eta_i \eta_r \eta_e$$

The external efficiency is the ratio of the optical power output (P_{out}) to the input electric power (P_{in}).

$$\eta_0 = \frac{P_{out}}{P_{in}}$$

For most of the common LEDs, η_0 ranges from 0.01 to 0.05.

3.8.1 Injection Efficiency

This gives how effectively the minority charge carriers are injected. Under forward bias. In n^+-p system, the electron injection is more dominant than the hole injection. The injection efficiency in this system is the fraction the electron current being injected into the p-side to the total current.

$$\eta_i = \frac{\text{Electron current density}}{\text{Total current density}}$$

$$= \frac{J_e}{J_e + J_{ph}}$$

$$= \left(1 + \frac{\mu_h p_p L_e}{\mu_e n_n L_h}\right)^{-1}$$

where $L_{e(h)}$ are the diffusion lengths, p_p the acceptor doping in the p-side, n_n the donor doping on the n-side, μ_e, μ_h mobility of electrons and holes in a III–IV system.

$L_e \approx L_h$ and $\mu_e \gg \mu_h$ then for $n_n \gg p_p$; the injection efficiency $\eta_i \approx 1$.

3.8.2 Radiative Recombination Efficiency

The radiative recombination efficiency is the ratio of the number of photons generated to the number of injected electron. LEDs with direct band gap materials it is about 50% for homojunction and about 80% for double hetrojunctions. In indirect band gap, the radiative recombination is very low, due to the non-radiative processes such as Auger recombination SRH recombination, recombination via surface states etc.

3.8.3 Extraction Efficiency

This factor is related to how efficiently the light is extracted from the LED, minimizing the losses due to different mechanisms. The extraction efficiency can be enhanced by minimizing the loss of light caused by following mechanisms.

Absorption within the Material

This is caused by the reabsorption of the emitted light by the material and depends on the absorption coefficient of the material. For minimizing the reabsorption, the emission region must be small. This loss can be reduced by placing the light emitting junction close to the surface in a direct band gap device. However, this will reduce the radiative recombination efficiency as the junction approach to the surface. The use of hetrojunction eliminates this problem to a large extent.

Fresnel Loss

When the light passes from a medium with index of refraction n_2 to a medium of refractive index n_1 a portion of light will be reflected back. This loss of light is called Fresnel loss. The reflection coefficient for normal incidence is

$$R = \left(\frac{n_2 - n_1}{n_1 + n_2}\right)^{1/2}$$

Critical Angle Loss

The third loss mechanism is caused by the total internal reflection of the photons incident to the angles greater than the critical angle

$$\theta_c = \sin^{-1}\left(\frac{n_1}{n_2}\right)$$

Light originating at recombination centres near the *p-n* junction will be radiated isotropically, however those with angles greater than θ_c are reflected back into first medium.

The transmission can be increased by employing proper radiation geometry. For example, as shown in Figure 3.8, in dome LED, a dome-shaped dielectric encapsulation with a transparent material of high refractive index is used.

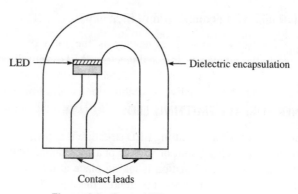

Figure 3.8 Dome LED.

3.9
LED MATERIAL CHOICE

Important requirements of a semiconductors to use it as a LED material are as follows:

(i) Emission wavelength is determined by the band gap of the semiconductor material, so that the material should have appropriate band gap to give the emission in the required wavelength range.

(ii) Both n and p type material should have the ability to dope at high level and also should have low resistivity.

For LEDs emitting in the IR and FIR range, low band gap materials are needed, and large band gap material have to be used for visible LEDs. However the difficulty is that the higher band gap materials have higher melting point, or lower in structural stability, They also have higher resistivity and are difficult to dope heavily.

Commonly used LED material and the wavelength range are listed as follows:

Material	λ
GaAs	860 nm
GaAs (Si doped)	910–1020 nm
GaP	550 nm
GaP (N doped)	590 nm
GaInP (N)	640–680 nm
GaAsP (N)	650 nm
GaAlAs (Si)	800–900 nm
GaAlAs (Zn)	650 nm

3.10
LED CONSTRUCTION

A *p-n* junction formed with a single semiconductor is called a homojunction. A junction formed by dissimilar semiconductors of unequal band gaps is called a hetrojunction. If it contains two hetrojunctions it is called a double hetrojunction (DH). High performance optoelectronic devices are constructed by hetrojunction structure.

3.11
PLANER SURFACE EMITTING LED

A typical planer surface emitting LED structure is shown in Figure 3.9. The structure is formed by growing liquid or vapour phase epitaxial method. It is designed such that the radiative recombination is taking place at the side of the junction near the surface, where the re-absorption is greatly reduced.

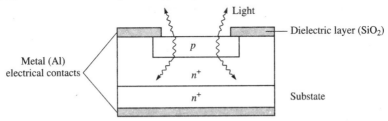

Figure 3.9 Surface-emitting LED.

3.12
DOUBLE HETROJUNCTION (DH) LED STRUCTURE

In homojunction light emitting diodes made from a single semiconductor material, the internal quantum efficiency of the device will reduce due to the following reasons:

(i) For the emitted photons not to absorb the photon emission volume must be close to the surface. Near the surface the semiconductor quality is usually not very good due to the presence of defect states. These surface states on the p-layer close to the light emitting junction give rise to non-radiative recombinations. If the surface is far away from the junction, the reabsorption probability will be enhanced.

(ii) The electrons injected from the n^+ side into the p-region, diffuse over a long distance before recombining with holes, thereby enhancing the reabsorption. This reduces the internal quantum efficiency.

In a DH LED, these problems are reduced by injecting charges froms a larger band gap material into a narrow band gap thin active region. A schematic diagram of DH light emitting diode is shown in Figure 3.10. The electrons and holes are injected from the top n^+ and p layers of AlGaAs, respectively into a thin GaAs active region. Such hetrojunction LEDs are formed by epitaxial growth with active region of width 0.1 to 0.2 μm. In these, the efficiency is increased because of higher injection efficiency, reduced non-radiative recombination and reduced reabsorption.

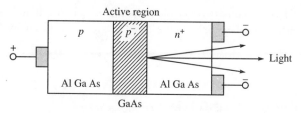

Figure 3.10 DH structure LED.

3.13
EDGE EMITTING LED

In an edge emitting LED, the active layer is a lightly doped thin layer of $n - \text{In}_{0.47}\text{Ga}_{0.53}\text{As}$ in between n^+ InGaAsP and p^+ InGaAsP layers. There is also an optical cladding layer of InP at the outerside of InGaAsP layers. The metal stripe contacts restrict the charge carriers in the lateral direction. Under a forward bias injection a large number of charge carriers recombine in the active region. InGaAsP layers on both side of the active layer acts as a waveguide. InP cladding confine the electron and holes in the active region and also cause the photons to travel along the LED axis. The photons generated in the active region spread into the guiding layer without reabsorption and emerge from the edge of the device. The emitted region is in rectangular shape with a thickness of a few micron and width about 60 microns and gives a more collimated light emission.

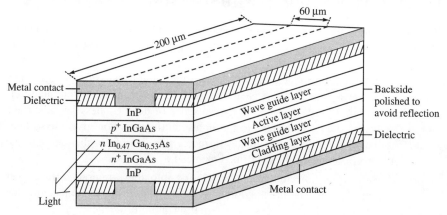

Figure 3.11 Edge emitting LED.

EXAMPLE 3.1 Compute the responsivity of an LED emitting at 600 nm, if its efficiency is 3%.

Solution
$$R = \eta \frac{h\nu}{e}$$
$$= 0.03 \frac{6.6 \times 10^{-34} \times 3 \times 10^8}{600 \times 10^{-9} \times 1.6 \times 10^{-19}} = 0.06 \text{ W/A}$$

Light Emitting Diode — 47

EXAMPLE 3.2 A planner LED exhibits an external power efficiency of 1% when driven with a current 40 mA. The voltage across its terminal is 2.2 volts. Estimate the optical power generated.

Solution
$$\eta = \frac{\text{Optical power}}{\text{Input electrical power}}$$
$$= \frac{P}{IV}$$
$\therefore$
$$P = 0.01 \times 40 \times 10^{-3} \times 2.2$$
$$= 0.88 \text{ mW}$$

EXAMPLE 3.3 The efficiency of a light emitting diode operating at 50 mA is 2%. The bandgap of the material is 1.4 eV. Calculate the optical power that can be extracted from it.

Solution
$$P = \eta I \frac{h\nu}{e}$$
$$= 0.02 \times 50 \times 10^{-3} \times 1.4$$
$$= 1.4 \text{ mW}$$

EXAMPLE 3.4 Show that, $\lambda = \dfrac{1.24}{E_g}$

where E_g is the bandgap in electron volt and λ the wavelength in microns.

Solution
$$\frac{hc}{\lambda} = E_g \quad (\text{in joule})$$
$\therefore$
$$\lambda = \frac{hc}{E_g}$$
$$= \frac{6.6 \times 10^{-34} \times 3 \times 10^8}{E_g(\text{eV}) \times 1.6 \times 10^{-19}}$$
$$= \frac{1.24}{E_g(\text{eV})} \; (\mu m).$$

REVIEW QUESTIONS

1. Why silicon and germanium are not used for making LEDs?
2. Define radiative efficiency of LED.
3. Describe edge emitting LED.
4. Define spectral response of an LED.
5. Discuss the efficiency of LED with suitable theory.
6. Write short notes on
   Injection luminescence
   Photoluminescence
   Cathodoluminescence

7. Explain injection luminescence in LED.
8. What are the dominant radiative recombination mechanism in LED?
9. Explain the performance characteristics of LED?
10. Write short notes on
    Injection efficiency
    Recombination efficiency
    Extraction efficiency
    External efficiency.
11. Explain the choice of LED materials.
12. Explain the responsivity of a LED.
13. Explain the structure and working of a DH LED.
14. Describe planner surface emitting LED.

4

Semiconductor Lasers

4.1
INTRODUCTION

Semiconductor lasers are similar to other lasers in which the emitted radiation has spatial and temporal coherence, high monochromaticity and directionality. However, semiconductor lasers differ from other lasers in several aspects as follows:

(i) In conventional lasers, the quantum transition occur between discrete energy levels, whereas in semiconductor lasers, the transitions are associated with the band properties of the semiconductor. As a result, semiconductor laser spectral lines are much broader than the conventional lasers.

(ii) The spectral and spatial characteristics of a semiconductor laser are strongly influenced by the junction properties; such as band gap, refractive index, doping level etc.

(iii) The laser action is produced by passing a forward current through the junction.

(iv) The active region is very narrow ($\approx$ µm). The beam divergence is considerably larger than that of conventional lasers.

(v) A semiconductor laser is very compact in size (~0.1 mm)

(vi) Semiconductor lasers have short photon lifetimes, so that modulation at high frequencies can be achieved.

Because of its compactness and capability for high frequency modulation, the semiconductor diode laser (SDL) is one of the most important source for optical fibre communication.

4.2
DIFFERENCE BETWEEN LED AND SDL LIGHT EMISSION

The light emitted in LED is by spontaneous emission, whereas it is by stimulated emission is SDL. Therefore, in addition to the electron injection provided by forward bias, the requirements for stimulated emission and the optical feedback to compensate the losses have to be provided in the laser diode.

The spontaneous emission process requires a smaller forward bias, and therefore, LEDs operates at lower current densities then SDLs.

The emitted photons have random phases in LED, whereas it is coherent in SDL.

The line width of the emitted light is typically 30–40 nm in LED. The line width of the optical spectrum of a diode laser is very narrow.

4.3
LASER ACTION IN A p-n JUNCTION

When a forward bias is applied to a semiconductor diode laser, a large number of electrons and holes are injected into the opposite side of the junction. This result into a non-equilibrium carrier concentration in each side, i.e. under forward biasing the density of minority carriers is larger than equilibrium or attain a state of population inversion. This occurs over a small region about a diffusion length where the electron and holes recombine radiatively. This layer is known as active region. The photons emitted by radiative recombination can stimulate a downward electron transition emitting photons of same phase and frequency. If the gains due to stimulated emission is sufficiently larger than losses in the medium, coherent radiation—laser—will emit through the active region.

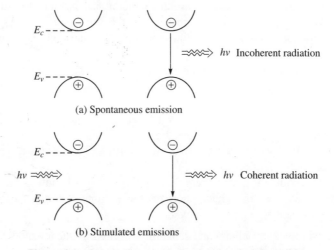

Figure 4.1 Spontaneous and stimulated emissions in SDL.

4.4
CONDITIONS FOR LASER ACTION

We have seen that under forward biasing a non-equilibrium of charge carriers is created in the SDL. The recombination of these charge carriers stimulate further recombination and the selective amplification result into coherent radiation.

Two necessary condition for laser action are:
(i) The gain due to stimulated emission must be sufficiently larger than (or at least equal to) the losses in the medium.
(ii) The emitted radiation must be coherent.

In order to attain these conditions, the laser medium must be placed in a resonant cavity. Usually a Fabry-Perot cavity structure is used, in which the two ends are cleaved as optically flat mirrors so as to form a waveguide. The spontaneously emitted light makes a round trip in between the mirrors, so that the feedback provided selectively amplify one of the frequency corresponding to the peak of the spontaneous output. Laser action occurs when the gain provided by the medium overcomes the loss in the cavity for a round trip.

4.5
THRESHOLD GAIN

The threshold condition necessary for laser action is that, the gain must be at least equal to the losses in the medium. Consider a Fabry-Perot cavity formed with two mirrors of reflectance R_1 and R_2, separated by a distance L (Figure 4.2).

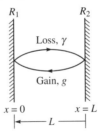

Figure 4.2 Fabry–Perot laser cavity.

Let g and γ be the gain and loss coefficients respectively. The initial intensity, is represented as I_0, when this is reflected by R_1 and as it reached at the second mirror the intensity is

$$I_1 = I_0 R_1 \exp(g - \gamma)L \tag{4.1}$$

As reflected by R_2 and after travelling a distance L towards the first, the intensity becomes

$$I = I_2 = I_1 R_2 \exp(g - \gamma)L \tag{4.2}$$
$$= I_0 R_1 R_2 \exp 2(g - \gamma)L \tag{4.3}$$

In a lasing medium $g > \gamma$, the intensity grows hence there is net amplification. Equation (4.1) gives the intensity after travelling a distance $2L$ in the cavity. At the threshold, the round trip gain is equal to the losses, i.e., $I = I_0$.

i.e.
$$R_1 R_2 \exp 2(g_{th} - \gamma)L = 1$$

which gives the threshold condition for laser action. The threshold gain

$$g_{th} = \frac{1}{2L} \log_e \left(\frac{1}{R_1 R_2}\right) + \gamma$$

Initial gain is much larger than threshold gain and after a few oscillations it reaches a steady value.

The losses in the cavity happens due to many reasons. Some of them are as follows:

(i) transmission at the mirrors
(ii) absorption by the cladding
(iii) absorption, scattering and diffraction at the mirrors
(iv) presence of defects impurities, etc.

The net loss coefficient can be expressed as with $R_1 = R_2$

$$\gamma = \frac{1}{L} \log_e R$$

4.6
POWER OUTPUT FROM A SDL

The output power of a semiconductor diode laser can be expressed as

$$P_0 = A(J - J_{th}) \left(\frac{\eta_i h\nu}{q}\right) \left[\frac{(1/2)L \log_e (1/R_1 R_2)}{\gamma + g_{th}}\right]$$

where η_i is the internal quantum efficiency and A is the area,

$$J = qdR_{sp} \quad (\text{Amp cm}^{-2})$$

is the current density in a SDL; d is the width of the active layer and R_{sp} is the total spontaneous recombination rate (cm^{-3}s^{-1}) J_{th} is the threshold current density.

The increase in laser output to the increase in the drive current is known as differential quantum efficiency given by

$$\eta_d = \eta_i \left[\frac{\log_e (1/R)}{\gamma L + \log(1/R)}\right]$$

The overall quantum efficiency which is related to the power conversion efficiency, is defined as

$$\eta_p = \frac{P_0}{V_f AJ}$$

where V_f is the forward bias voltage applied. The overall efficiency of a SDL depends on the pumping efficiency, the gain factor g and the losses in the cavity γ.

4.7
BASIC STRUCTURE OF A JUNCTION LASER

Figure 4.3 shows the basic structure of a *p-n* junction laser. The active medium is a junction formed by a *p*-type and *n*-type semiconductors. The structure is designed

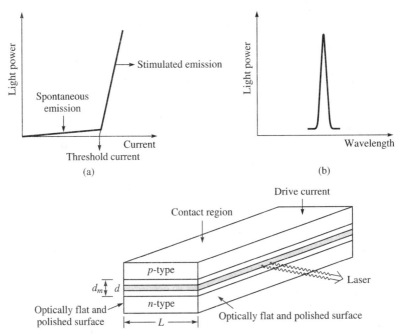

Figure 4.3 (a) Current-light power characteristics of SDL, (b) Spectral response of SDL, (c) Fabry-Perot SDL cavity structure.

as an optical cavity which can amplify the photons generated. For semiconductor lasers the most widely used cavity structure is Fabry-Perot cavity in which a pair of plane surface are cleaved at the two ends to provide required reflectance perpendicular to the plane of the junction. The remaining sides are roughened to eliminate lasing from these directions.

When a forward bias is applied to the SDL, a current flows initially at low level, there is spontaneous emission in all direction. As the bias is increased further the stimulated emission occur and a monochromatic and highly directional beam of light is emitted from the junction. The injection current at which this occurs is called the threshold current. The spectral output at the threshold is shown in the Figure 4.3.

Population inversion or gain occurs over an active region of thickness d which is about the diffusion length of the injected carrier. Within the active region the gain will eventually exceed the losses with an enhancement in the rate of stimulated emission. Outside the active region the losses will dominate. The thickness over which the optical mode extends is larger than the diffusion length d. As the rate of stimulated emission increases in the active region, the round trip gain in the cavity overcome the losses and laser commences.

The resonant modes satisfy the conditions

$$L = \frac{m\lambda}{2}$$

where, m is the mode number 1, 2, ..., etc, λ is the wavelength and L is the cavity length.

4.8
HETROJUNCTION LASERS

In a homojunction laser the threshold current density is very large. Moreover the active region and the thickness of the mode volume is not well-defined so that the carrier and optical confinement are very poor. Such homojunction lasers cannot be operated continuously for a long time at room temperature. By using a hetrojunction structure, the threshold current can be reduced and also the carrier and optical mode confinement can be achieved.

4.8.1 Single Hetrostructure (SH) Laser

In SH laser, a layer of p-GaAs is sandwitched between n^+ GaAs and p-GaAlAs layers. p-GaAlAs has a larger band gap and has a lower refractive index than GaAs. The electrons injected from the n^+-GaAs to p-GaAs is blocked by higher band gap p-GaAlAs, and hence, confine the carriers in the active region. Since the refractive index of p-GaAlAs and n^+ GaAs is smaller than p-GaAs the optical modes are confined in the active region. The active region thickness is same as that of the p-GaAs region. If its thickness is made small the same carrier injection level can be attained with a small drive current which will improve the power efficiency considerably.

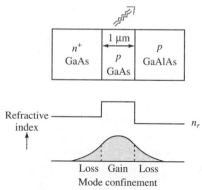

Figure 4.4 SH hetrology structure.

4.8.2 Double Hetrostructure (DH) Laser

In a DH laser the active region is a thin layer of p-GaAs in between GaAlAs layers. The higher band gap of GaAlAs confine the carriers, and its lower refractive index confine the optical modes within the active region. Thus, in a DH laser the carrier and optical confinement are achieved simultaneously. The diffusion length lies about 0.1 to 0.3 μm. The threshold current is also small.

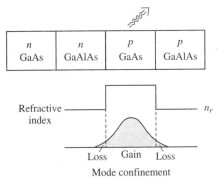

Figure 4.5 DH laser.

4.8.3 Stripe Geometry Hetrostructure

A stripe geometry hetrostructure is formed by different ways. They are mesa etching, proton implantation and mesa etching and regrowth. The important and advantages of stripe geometry are as follows:

(i) The output is usually single mode
(ii) The output light can be easily coupled to optical fibres
(iii) The non-linearities in the current-light power characteristics are eliminated.

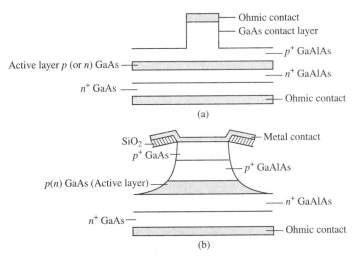

Figure 4.6 (a) Stripe geometry hetrostructure formed by mesa etching. (b) Stripe geometry hetrostructure formed by mesa etching and regrowth of n GaAlAs.

4.9
QUANTUM WELL LASERS

A Quantum Well (QW) laser is similar to DH laser except the active layer is much thinner about 5 to 10 nm. Due to the this active layer the threshold current in low. Here the electrons and holes are spread over a small energy range with a relatively high density at the band edges, so that the population inversion can be easily attained. Schematic of a layered structure of a QW laser is shown in Figure 4.7. Because of the extreme narrowness of the gain region, the optical confinement is very poor. The multiple QW structure allow higher optical confinement and can be operated at very low threshold current of a few milliamperes.

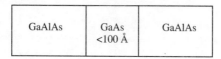

Figure 4.7 QW laser structure.

4.10
DISTRIBUTED FEEDBACK LASER

Semiconductor lasers with Fabri-Perot cavity are formed by cleaving and polishing the cavity facets to provide necessary feedback for laser action. It is easy to fabricate, however it has a drawback that, there are several preference given to a particular mode. The mode domination is determined by the electronic properties of the active region.

An important approach towards the design of a mode selective optical cavity is using corrugated structure or grating (Figure 4.8).

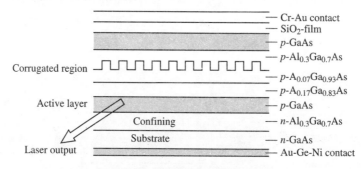

Figure 4.8 Schematic of DFB laser.

In such a structure a periodic variation of the refractive index within the cavity along the direction of wave propagation is produced. Feedback in such a cavity occurs due to the energy of the wave propagating in the forward direction by Bragg's diffraction at the corrugation or grating. Lasers that uses these corrugated structure are called Distributed Feedback Lasers (DFB lasers) or Distributed Bragg Reflector (DBR) lasers. The fabrication process involves growing the basic laser structure, etching a periodic structure and regrowing the top layer. The grating structure should be as close as possible to the active region so that the optical wave interacts strongly with the grating.

REVIEW QUESTIONS

1. Compare the properties of a semiconductor laser with conventional laser.
2. What are the differences between LED and SDL?
3. Explain the principle of laser action in a *p-n* junction laser.
4. Discuss the conditions of laser action in a SDL.
5. What are the losses in a laser cavity?
6. Explain the following laser structure.
   (a) homojunction (Fabry-Cavity structure)
   (b) SH and DH SDL
   (c) Stripe geometry structure
   (d) DFB laser
   (e) QW laser

5

Photodetectors

5.1
INTRODUCTION

Electronic devices commonly used for the detection of optical radiation are called optoelectronic detectors. These detectors convert photon flux into electrical signal which is then amplified with a proper electronic circuit. Commonly used photodetectors are

1. Photoconductive detectors
2. Photodiodes
3. PIN diodes
4. Avalanche photodiodes
5. Photomultiplier tubes
6. Charged coupled devices

Sensitivity of these photodetectors is very high. Some detectors such as photomultipliers are capable of detecting even a single photon. Two important aspects that are commonly used to specify photodetectors are their response time and the spectral responsivity. Response time is the minimum time taken by a detector to respond the optical field. They have very short response time, detectors with pico second response time is now available. The spectral responsivity is the wavelength range over which they respond. Depending on the materials used, the spectral responsivity varies from ultra violet to far infrared region.

The basic steps involved in a photodetection process are
(i) Absorption of photons by materials generating charge carriers.
(ii) The transportation of these charge carriers.
(iii) Electrical circuitary to measure the photogenerated current.

The performance requirements for the detectors are high sensitivity, low noise, fast response, wide band width, high reliability and low cost.

5.2
PHOTOCONDUCTIVE DETECTORS

The photoconductive detectors are basically semiconductors—n or p type—That are appropriately doped; n-type material is a donor impurity doped semiconductor, whereas *p*-type material is an acceptor impurity doped semiconductor. A voltage is applied across the detector through a load resistance as shown in Figure 5.1.

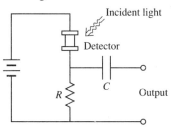

Figure 5.1 Photoconductive detector circuit.

With the bias voltage applied the resistance of the photoconductor without any incident photon flux is called the dark resistance which is much larger than the load resistance and the current flowing is called the dark current. By illuminating the detector with photons of sufficient energy, electrons may be excited into the conduction band in the *n*-type semiconductor, or holes may be created in the valence band of the *p*-type semiconductor.

With $E_g = E_e - E_v$, the bandgap energy of the material, and v, the photon frequency, then if,

$$h\nu \geq E_g$$

or

$$\lambda \leq \frac{hc}{E_g}$$

the electrons are created in the conduction band, and will cause an increase in the conductivity of the semiconductor. This phenomenon is known as photoconductivity which is the basic mechanism operative in photoconductive detectors.

The incident light creates a large number of carriers in the detector chip and reduces its resistance, thereby causes an increase in the voltage across the load resistance. Once the electron (or hole) has been created by the absorption of a photon it will drift under the influence of the electric field towards the appropriate contact, and will generate a current in the circuit. As long as photon absorption and photo excited ionization continues, the current flows in the external circuit.

If $N = (n/t)$ is the number of photons absorbed per second by the semiconductor, the optical power absorbed is given by

$$P = Nh\nu$$

where v is the frequency of the radiation. The quantum efficiency is defined as the ratio of the rate of generation of photo excited charge carriers to the rate of photons absorbed, i.e.

$$\eta = \frac{\text{Rate of generation of charge carriers}}{\text{Rate of photon absorption}}$$

$$= \frac{\text{No. of charge carriers generated per time}}{\text{No. of photons absorbed per time}}$$

$$\eta = \frac{G}{N}$$

∴ The rate of generation of charge carriers

$$G = \eta \frac{P}{h\nu}$$

If the lifetime of a carrier is τ_0 and the time taken by it to cross the detector is τ_d, the photocurrent generated in the external circuit is

$$i = \eta \frac{e}{h\nu} \left(\frac{\tau_0}{\tau_d} \right) P$$

where (τ_0/τ_d) is known as the photoconductive gain. The photocurrent is proportional to the square of the electric field of the light since $i \propto P \propto E^2$. Detectors with such characteristics are called square law detectors. The response time of photoconductive detectors is typically in the micro second range. The photon energy required to excite electrons into the conduction band (or to create holes in the valence band) depends on the band gap of the semiconductor and varies form material to material.

The capability to detect long wavelength radiation up to 30 µm is the principal advantage of photoconductive detectors over other photodetectors. However, due to very low energy requirement for creating photoelectrons at room temperature noise current may be produced. To avoid such thermal noise most of the photoconductive detectors working in the far infrared region has to be cooled to liquid nitrogen (77 K) (or even to liquid helium 4.2 K) temperature and consequently are bulky and complicated in structure.

5.3
PHOTOCONDUCTIVE MATERIALS

5.3.1 Doped Semiconductors

Typical doped semiconductor photoconductive detectors are zinc and boron doped germanium detectors which have spectral response from 20 µm to 100 µm. Cooling is required to reduce background noise. They have response time around a few microseconds.

5.3.2 Cadmium Sulphide and Cadmium Selenide (CdS and CdSe)

Both are used as low cost visible radiation sensors which is commonly used in light meters, cameras, etc. They have a response time of about 50 ms. A typical geometry of a CdS photoconductive cell is shown in Figure 5.2. A film of the material in

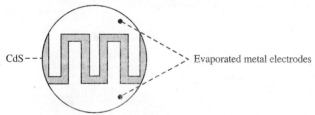

Figure 5.2 CdS photoconductive detectors.

polycrystalline form is deposited on an insulating substrate. The electrodes are formed by evaporating a suitable metal such as gold through a mask to give a high sensitivity to the device.

5.3.3 Lead Sulphide (PbS)

Lead sulphide (PbS) is a near infrared photodetector material with a wavelength response from 1 to 3.4 μm and have a response time about 200 μs. The spectral range can be extended upto 4 μm by cooling to –30°C.

5.3.4 Indium Antimonide (InSb)

They have spectral range extending up to 7 μm and exhibit response time around 5 ns.

5.3.5 Mercury Cadmium Telluride (HgCdTe)

It is an alloy of semimetal HgTe and semiconductor CdTe which have a spectral range in the 5–14 μm region.

5.4
JUNCTION PHOTODIODE

5.4.1 *p-n* Junction

A *p-n* junction is formed by bringing a piece of *p*-type and *n*-type semiconductor together. *n*-type material is a donor doped semiconductor in which the charge carriers are dominantly electrons and *p*-type material is an acceptor doped semiconductor in which the charge carriers are holes. The abrupt transition from the *n*-region to the *p*-region of two materials is known as *p-n* junction.

Figure 5.3 *p-n* junction.

When these two pieces are put into contact, electrons flow from *n*-region to the *p*-region leaving behind the immobile donor ions. Similarly holes flow from *p* to *n* region leaving the negatively charged acceptors behind. These immobile donors and acceptors built upon internal electric field around the junction which create a potential barrier that prevents further migration of carriers. When a hole in the *n*-region wanders in the vicinity of the *p-n* junction it is immediately accelerated by the internal field and drift to the *p*-side. Similarly, if an electron in the *p*-region wanders in a position close to the junction, it is accelerated by the internal field and drift into the *n*-region. Both processes contribute to the drift current, which is balanced by an opposite current due to the diffusion of majority carriers against the retarding field.

When a reverse bias is applied by an external circuit, almost all the majority mobile carriers in an area close to the *p-n* junction are swept by the bias electric field with only the immobile donor and acceptor behind as in Figure 5.4. Such an area is called depletion area. The increased internal electrical field built up in the depletion area form a potential barrier that greatly reduces the diffusion of majority carriers across the junction. There are always some minority carriers-electrons in the *p*-region and holes in the *n*-regions that happens to move in the vicinity of the depletion area and drift further into the other side of the *p-n* junction. The drift current under reverse bias is very small generally being if the order of microamperes.

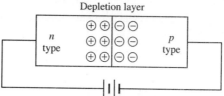

Figure 5.4 *p-n* junction with reverse bias.

5.4.2 Photodiode: Principle

If a reverse biased *p-n* junction is illuminated, the current varies almost linearly with the light flux. This effect is exploited in the semiconductor photodiode.

When a photon is absorbed in the *p*-region an electron-hole pair is created. The electron as a minority carrier will drift towards the depletion layer across the junction and contribute a charge to the external circuit. Similarly, a hole created by photoabsorption in the *n*-region will drift in the opposite direction and also contribute to the current. It should be noted that only those electron-hole pairs generated within the depletion area or within a diffusion length of the depletion area contribute to the current flow in the external circuit. Diffusion length means that the average distance that a minority carrier traverses before recombine with a majority carrier. Those electron-hole pairs created deep in the neutral region, i.e. the region far away from the depletion layer—will recombine with carriers of opposite type before they diffuse to the junction and thus do not contribute to the current flow.

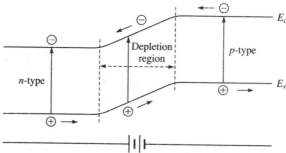

Figure 5.5 Formation and of charge carries in a reversed *p-n* junction under photoabsorption.

5.4.3 *V-I* Characteristics of a Photodiode

Typical volt-ampere characteristics of a photodiode is shown in Figure 5.6. When a reverse bias is applied an almost constant current independent of the magnitude of the reverse bias is obtained. The dark current corresponds to the reverse saturation due to the thermally generated minority carriers.

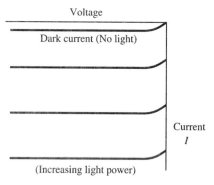

Figure 5.6 V-I characteristics of a photodiode at various light power.

5.4.4 Design of Photodiodes

The most common semiconductor material used for photodiode is silicon. It has a band gap energy of 1.4 eV and quantum efficiency up to 80% in the wavelength range 0.8–0.9 nm. A typical construction is shown in Figure 5.7. A junction is formed between heavily doped p-type material (p^+) and fairly doped n-type material, the depletion region extends well into the n-type material. The p^+ layer is made thin. The electrical contacts to the semiconductors material n^+ or p^+ is made via a metal. The detection efficiency may be increased by providing anti-reflection coating on the front surface of the detector by $\lambda/2$ coating of SiO_2.

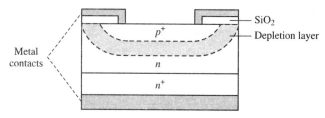

Figure 5.7 Design of a photodiode.

The radiation is allowed to fall upon the surface across the junction. The remaining sides are either painted black or enclosed in a metallic case. It is also possible to make detectors that are illuminated from the sides, i.e. parallel to the junction. This type of construction results in good sensitivity for wavelength close to the band gap limit where the optical absorption coefficient is relatively small.

5.4.5 Equivalent Circuit of a Photodiode

There are two different modes of operation available, namely photovoltaic and photoconductive mode. In the photovoltaic mode electrons and holes generated by the absorption of the incident light are collected at either end of the junction leading to a potential difference and a current flows if the device is loaded. In this mode the diode is unbiased. However, in the photoconductive mode, the diode is reverse biased and the electron-hole pairs generated by light absorption are separated by the high electric field in the depletion region. Such a drift of carriers induce a photocurrent in the circuit. If we assume a quantum efficiency η and if all the incident light being absorbed within the cell, then

$$i_\lambda = \eta \frac{I_0 A e \lambda}{hc}$$

where, I_0 is the irradiance falling on the cell of area A.

The equivalent circuit of a photodiode is shown in Figure 5.8. The photodiode is represented by a constant current generator i_λ-photocurrent generated by the light absorption with an ideal diode across it. The internal characteristics of the diode are represented by a shunt resistor (R_{sh}), a shunt capacitor (C_d) and series resistance (R_s). In the photoconductive mode, the diode is connected to a load, and a reverse voltage.

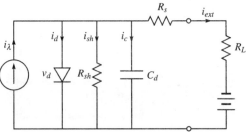

Figure 5.8 Equivalent circuit of a photodiode.

If the operation is at a fairly low optical modulation frequency, the effect of shunt capacitance can be neglected, i.e. $i_c = 0$. Then

$$i_\lambda = i_d + i_{sh} + i_{ext}$$

The diode current is given by

$$i_d = i_0 \left[\exp\left(\frac{eV_d}{kT}\right) - 1 \right]$$

where, i_0 is the reverse bias leakage current the diode voltage.

$$V_d = i_{sh} R_{sh}$$

and

$$V_{ext} = V_d - i_{ext} R_s$$

In photoconductive mode, $i_0 \approx 10$ nA and since $R_{sh} \approx 100$ M and $V_d \approx 10$ V, $i_{sh} \approx 100$ nA and i_λ is of the order of microamperes.
Thus

$$i_{ext} = i_\lambda = \frac{\eta e \lambda A I_0}{hc}$$

Hence, in the photoconductive mode, the external current flowing is directly proportional to the incident light irradiance.

5.5
PIN PHOTODIODE

Although a photon may be absorbed anywhere in the *p*-type or *n*-type material, only those photons absorbed in the depletion area or within a diffusion length can contribute to the external photocurrent. As the diffusion length is much larger than the width of the depletion region, most of the useful absorption and electron-hole pair production takes place outside the depletion area, but within the depletion length. This substantially limits the response time of a *p-n* junction photodiode.

In order to improve the response time of a photodiode, the carriers must be generated where the field is large so that the charge transport is primarily due to fast drift rather than slow diffusion. This is accomplished by adding an intrinsic region in between *p* and *n* layers as shown in Figure 5.9. Because of their sandwitched structure, this type of diode is known as PIN photodiode.

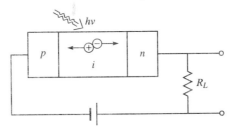

Figure 5.9 PIN photodiode.

In a PIN photodiode, the voltage drop occurs in the intrinsic layer. If the intrinsic layer is made wide enough such that most of the incident light is absorbed within it, the long diffusion time associated with the process that takes place outside the depletion area can be avoided. Therefore, PIN photodiodes respond much faster than the *p-n* junction photodiodes. Typically, the response time of a PIN photodiode is of the order of nanoseconds, whereas that of a *p-n* photodiode is of the order of microseconds. Since the intrinsic layer separates *p* and *n* layers, the capacitance of the device is greatly reduced. This further leads to a wider frequency response. Since the intrinsic region has no free charges, the resistance is high, and hence, most of the reverse voltage applied appears across the intrinsic region. The intrinsic region is usually wide so that the incoming photons have a greater probability of absorption in this region rather than in the *p* or *n* region. PIN photo diode is the most commonly used as solid state detector for faster radiation detection in the visible and infrared region.

5.5.1 Expression for Photocurrent

Consider an optical radiation of power P_0 incident on a PIN photodiode at $z = 0$. The unabsorbed radiation at any distance z is given by $P_0 \exp[-\alpha(\lambda)z]$, where $\alpha(\lambda)$ is the wavelength dependent absorption coefficient. The optical radiation absorbed in the semiconductor material over a distance z is then

$P(z)$ = Incident light power – Unabsorbed light power
$$= P_0 - P_0 \exp[-\alpha(\lambda)z]$$
$$= P_0 \{1 - \exp[-\alpha(\lambda)z]\}$$

An optical radiation of power P_0 incident on the photodetector first suffers a partial reflection at the air-semiconductor surface before entering the detector. Let R be the reflectivity, i.e. fraction of the incident light reflected. Then the optical power entering the semiconductor is $P_0(1 - R)$. Let w be the width of the depletion region, then the optical power absorbed in a distance w will be

$$P_0(1 - R)\{1 - \exp[-\alpha(\lambda)w]\}$$

If v is the frequency of the incident light, the number of photons absorbed per unit time

$$\frac{n}{t} = \frac{P_0}{hv}(1 - R)\{1 - \exp[-\alpha(\lambda)w]\}$$

Each absorbed photon leads to the generation of electron hole pairs. The quantum efficiency is given by

$$\eta = \frac{N/t}{n/t}$$

where N/t is the charge carriers per unit time.
Then

$$\frac{N}{t} = \eta \frac{n}{t}$$

$$= \eta \frac{P_0}{hv}(1 - R)\{1 - \exp[-\alpha(\lambda)w]\}$$

Assume that only a fraction δ of the charge carriers produced contribute to the photocurrent. The remaining e-h pairs decays by recombination. Then the photocurrent is given by

$$I = \frac{Ne}{t}\delta$$

$$= \eta \frac{e}{hv}(1 - R)\{1 - \exp[-\alpha(\lambda)w]\}\delta P_0$$

and the quantum efficiency is given by

$$\eta = \frac{I/e}{P_0/hv}$$

$$= (1 - R)\delta\{1 - \exp[-\alpha(\lambda)w]\}$$

The responsivity ρ is the photocurrent generated per unit optical power, i.e.

$$\rho = \frac{I}{P_0} = \frac{\eta e}{hv}$$

$$= \frac{e\delta}{hv}(1 - R)\{1 - \exp[-\alpha(\lambda)w]\}$$

Unit of δ is Ampere per watts.

Let R_L be the load resistance connected in series with the reversed bias PIN photodiode. In the absence of any light the entire bias voltage drop across R_L is zero. When the light with power P_0 falls on the photodiode, it leads to a current δP_0, so that the voltage drop across R_L is

$$V_R = \delta \rho P_0 R_L$$

The voltage V_R across the load is proportional to the optical power P_0 falling on the photodetector. Thus, the measurement of V_R gives the optical power P_0. When $V_R = V_b$, the reverse voltage, the maximum value of optical power measurable is

$$P_{max} = \frac{V_b}{\rho R_L}$$

Beyond this power the photodiode gets saturated. Thus, by choosing an appropriate value of R_L one can operate the PIN photodiode over an optical power range 0 to P_{max}. The sensitivity is given by

$$\frac{V_R}{P_0} = \rho R_L$$

Increasing R_L increases the sensitivity while reduces the range of detection.

5.5.2 Response Time

The response time and hence the bandwidth of a photodiode depends on three primary factors described as follows:

(i) The transist time of the photogenerated carriers through the depletion region is given by

$$\tau_t = \frac{w}{v_d}$$

where w is the depletion region width and V_d is the carrier drift velocity. τ_t is approximately the photodiode rise time. The requirement of smaller ω is contrary to that required to achieve larger quantum efficiency.

(ii) The electrical frequency response is determined by the RC time constant, which depends on the diode capacitance given by

$$C_d = \frac{\varepsilon A}{w}$$

where, ε is the permittivity of the semiconductor, A is the diode area and ω is the depletion layer width. The rise time is given by

$$t_r = 2.19 R_L \frac{\varepsilon A}{w}$$

To achieve small rise time, the area A and the resistance R_L must be small, and the width of the depletion region must be large. As w increases, the transist time will increase, so a balance has to be made with decreasing R_L. The bandwidth of the photodiode is given by

$$\Delta f = \frac{1}{2\pi R_L C_d}$$

(iii) Slow diffusion of carriers generated outside the depletion region. The time taken by carrier to diffuse a distance d is given by

$$\tau_{\text{diff}} = \frac{d^2}{2D_C}$$

where D_C is the minority carrier diffusion coefficient.

To achieve a small rise time and a larger bandwidth R_L should be small. To increase the quantum efficiency it is required to increase the thickness of the depletion region. This on the otherhand increases the response time and lower the bandwidth. Hence, a compromise has to be made in choosing the depletion width.

5.6

AVALANCHE PHOTODIODE

In a photodiode, each photon absorbed around the depletion area can create only one electron-hole pair, and therefore, no gain in the detection. However, by increasing the reverse bias voltage across the *p-n* junction, the field in the depletion area can be increased to a point at which the carriers-electrons or holes—that are accelerated across the depletion area can gain enough kinetic energy to stimulate new electrons from the valence band to the conduction band. These secondary generated charge carriers, are further accelerated under the influence of the field and create more electron-hole-pairs. This process leads to an avalanche multiplication of charge carriers and the photodiodes using this effect are called avalanche photodiode (APD)

Figure 5.10 describes the energy level diagram illustrating avalanche multiplication in an APD. The photon absorption in a reverse bias creates an electron-hole pair (AB). If this electron has sufficient kinetic energy to excite an electron from the valence band to the conduction band, it can create a new electron-hole pair. The newly generated carrier drift in opposite direction. The holes (F)

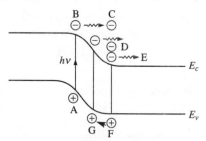

Figure 5.10 Avalanche multiplication in APD.

can also cause carrier multiplication (G). The result is a dramatic increase in the junction current that sets in when the electric field is high enough. For avalanche multiplication to take place the diode must be subjected to a large electric field. Thus, in APD, several tens of volts to several hundreds of reverse bias voltage is applied. Typical variation of the current gain with reverse bias voltage for an avalanche photodiode is shown in Figure 5.11.

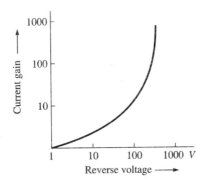

Figure 5.11 Current gain of an APD.

Current gain in excess of 100 are readily obtainable, but then the current gain is very sensitive to the reverse bias applied. If the bias voltage is made too large then a self-sustaining avalanche photocurrent flows in the absence of any photoexcitation, and this sets in upper limit to the voltage that can be used. To ensure a uniformly high avalanche breakdown voltage, it is necessary to provide uniform field across the device.

The two factors that limit the electron multiplication factors are as follows:

(i) Series resistance between the junction and the diode terminal.
(ii) As the multiplication starts the device temperature increases and this reduces the impact ionization coefficients for electrons and holes.

If M respresents the multiplication factor, the responsivity of the APD is given by

$$\rho_{APD} = \frac{M \eta e}{h \nu}$$

APD differ from PIN photodiode design in having an additional p-type layer between the intrinsic and highly doped n-region. The electron-hole pairs are still generated in the intrinsic region, but the avalanche multiplication takes place in the p-region. Photons absorbed in the i-region excite electron-hole pairs, and on acceleration, as a result of collision an avalanche multiplication of carriers occurs in the p-region.

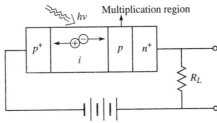

Figure 5.12 APD.

A number of design have been proposed. For example guard ring structure is shown in Figure 5.13. The guard ring restricts the avalanche region to be the

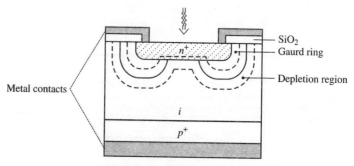

Figure 5.13 Schematic of APD.

central illuminated part. The guard ring is a region of comparatively low doping, and hence, the depletion region extend an appreciable distance into it. Thus, in the vicinity of the guard ring, the total depletion layer thickness is greater, and hence, the maximum electric field strength is less than the field strength in the central region.

As APD has high gain so that they are very suitable for the detection of light with low photon intensities. The device bandwidth is small and noisy. Frequency response is limited by diffusion, drift across the depletion layer and capacitative loading. The noise level depends on the carrier multiplication factor and impact ionization coefficient.

A number of precautions are necessary when using an APD. The rapid variation of gain with bias voltage requires the use of a very stable power supply to maintain a constant gain. Care must be taken if the simple bias circuit is used since the relatively large output current may cause significant voltage drop across the bias resistor, and the diode series resistance give rise to non-linearities in the output. A further difficulty is that the gain is highly dependent on temperature so that it needs temperature stabilization circuits.

5.7
MODULATED BARRIER PHOTODIODE

The modulated barrier photodiode is a device in which a barrier is placed in the $n^+ - n^-$ structure with a heavily doped p^+ layer. The thickness and doping of the p^+ layer is so choosen that it is fully depleted at the zero bias. The dominant conduction mechanism in the device is by thermionic emission over the barrier. There is significant barrier lowering with the reverse bias.

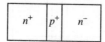

Figure 5.14 Modulated barrier photodiode.

They can be used as a very efficient photodetector with high optical gain. With its low noise and high gain at low incident light power, this detector play an important role in low signal level detection system.

5.8
METAL SEMICONDUCTOR PHOTODIODE (SCHOTTKY BARRIER PHOTODIODE)

Figure 5.15 shows schematic cross-section of a metal semiconductor photodiode fabricated on a semi-insulating substate. An n^+ substrate layer, n^- absorbing layer and a thin layer of semitransparent metal layer are deposited on it. A thin dielectric anti-reflection coating is deposited on the semitransparent layer and the top.

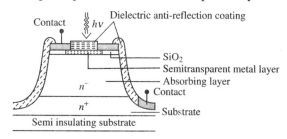

Figure 5.15 Metal semiconductor photodiode.

The metal semiconductor photodiode is operated under reverse bias. It has V-I characteristics very similar to that of ordinary p-n junction. Schottky photodiodes are unipolar devices, the charge carriers are majority carriers (electrons). The current is due to the thermionic emission of these carriers. Since it is a majority carrier device minority carrier storage and their removal problems do not exist and therefore higher bandwidth can be obtained. The temporal response and bandwidth of the device are determined by transist time of carriers through the depletion region and the external circuit parameters. Schottky photodiode with bandwidth 100 GHz and rise time about 1 ps are available.

5.9
PHOTOTRANSISTOR

A bipolar phototransistor together with its circuit model are shown in Figure 5.16. It differs from a conventional bipolar transistor in having a large base-collector junction as the light collecting element. The frequency response is limited by large collector capacitance. Typical response time is about 5 µs.

Light absorbed within the base region causes an emitter current to flow through the load resistance R_L, and thus, the signal voltage appears across it. The base is floated (i.e. no external connection is made to the base) and the base region is exposed to the incident radiation. The current flows from the junction in internally amplified.

For a transistor, the emitter current is

$$I_E = I_C + I_B$$

where I_C and I_B are collector and base currents, respectively. The total collector current is the sum of the reverse saturation current I_{CBO} (i.e. the leakage current due

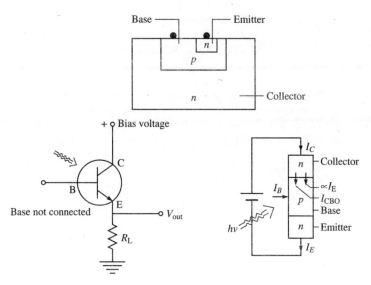

Figure 5.16 Circuit model of phototransistor.

to the motion of minority carriers across base-collector junction) and αI_E the part of the emitter current that reaches the collector terminal, where α is the common base current gain which is slightly less than unity. Thus,

$$I_E = I_{CBO} + \alpha I_E + I_B$$
$$= \left(\frac{1}{1-\alpha}\right)(I_B + I_{CBO})$$
$$= \left(\frac{\alpha}{1-\alpha}\right)(I_B + I_{CBO})$$
$$= (h_{fe} + 1)(I_B + I_{CBO})$$

Here, $h_{fe} = \alpha/(1-\alpha)$ is the common emitter current gain.

Assume that no radiation is falling on the phototransistor. Under this condition minority carriers are generated thermally and the electrons crossing from the collector to the base constitute the reverse saturation collector current I_{CBO}. With no incident radiation, base current $I_B = 0$ and the emitter current is dark current of the phototransistor given by

$$I_E = (h_{fe} + 1)I_{CBO}$$

If the light is turned on additional minority carriers are photogenerated, and this contribute to the base current. The photogenerated base current is thus

$$I_B = I_\lambda = \eta \frac{I_0 A e \lambda}{hc}$$

Thus, the external current flowing is

$$I_E = (I_\lambda + I_{CBO})(h_{fe} + 1)$$

and if $I_\lambda \gg I_{CBO}$, then

$$I_E = I_\lambda(1 + h_{fe})$$
$$= \eta \frac{I_0 Ae\lambda}{hc}(1 + h_{fe})$$

Figure 5.17 shows the V.I. Characteristics of a *npn* phototransistor for different light intensities.

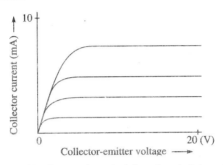

Figure 5.17 Output characteristics of a phototransistor.

Two main disadvantages of a phototransistor are
  (i) At low light levels the base current and h_{fe} can drop to quite low values.
  (ii) The frequency response is relatively poor (<200 kHz) mainly because of the time constant associated with capacitance and the resistance of the base junction.

5.10
MICROCAVITY PHOTODIODE

Two basic quantities that is required for high speed photodiode detectors are a large quantum efficiency and a large bandwidth. To increase the quantum efficiency it is required to increase the thickness of the depletion region, whereas a large bandwidth needs a thin depletion region. Such a situations is eliminated in a microcavity photodiode (Figure 5.18).

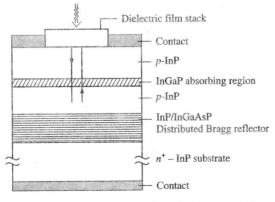

Figure 5.18 Schematic of a microcavity photodiode.

A thin low band gap absorption region is placed in the middle of a resonant cavity formed by heavily doped wide band gap region, and reflecting mirrors. The cavity is made with InP material with two mirrors. The top mirror is dielectric film stack. The bottom reflecting mirror is formed by a $\lambda/4$ superlattice distributed Bragg reflector. Reflectance close to unity is achieved with InP/InGaAsP superlattices. This cavity structure is grown on n^+ InP substracte provided with metal contacts. A resonance is builtup in the InP cavity at those frequency components of the incoming light for which the round trip phase shift is a multiple of 2π. Therefore, at resonance frequency the incoming light is reflected by two mirrors and round trip path length is greatly increased. The microcavity photodiodes make the simultaneous achievement of large bandwidth and high quantum efficiency.

5.11
PHOTOMULTIPLIER TUBE

5.11.1 Photoelectric Effect

Consider an electromagnetic radiation with a frequency v incident on a metal surface. If the incident photon energy hv is higher than the surface workfunction of the cathode material (W), electrons can be emitted with a kinetic energy

$$\frac{1}{2}mv^2 = hv - W$$

This is known as the photoelectric effect. No electrons will be emitted if

$$\frac{hc}{\lambda} < W$$

The maximum wavelength is known as the threshold wavelength and

$$W = hv_0$$

is called the threshold energy which is equal to the work function of the material.

Figure 5.19 shows a simple experimental set-up to observe photoelectric effect. In a vacuum tube two electrodes—a cathode and an anode—are placed a few centimetre apart. The anode is positively biased with respect to the cathode. When the cathode is illuminated with light such that the $hv_0 > W$, photoelectrons will be emitted from it which are collected by the anode. This constitute a current in the external circuit.

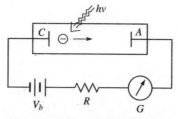

Figure 5.19 Photoelectric effect.

5.11.2 Photomultiplier Tube (PMT)

Photomuliplier tube is one of the most sensitive and common optical instrument used to detect and measure radiation in the near UV, visible and near IR region. It consists of a photocathode (C) a number of electrodes called dynodes (D) and an anode (A). The whole assembly is placed in a highly evacuated envelope in order to reduce the possibility of electron collisions with the gas molecules. The dynodes are maintained progressively higher potential with respect to the cathode (with about 100 V potential difference between the adjacent dynodes). A schematic diagram of a common PMT with linear cathode, dynode and anode assembly is shown in the Figure 5.20.

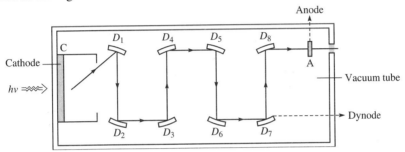

Figure 5.20 Photomultiplier tubes.

The dynode biasing is provided by means of a circuit as shown in Figure 5.21. A high voltage is applied through a resistor chain R_C, R_{D1}, R_{D2} ... etc., which act as a potential divider to maintain the dynodes D_1, D_2, D_3, ... etc., at increasingly higher positive potential with respect to the cathode.

When photocathode of the PMT is illuminated with light of wavelength $\lambda < \lambda_{max} = (W/hc)$, it converts the incident optical radiation into photocurrent by generating photoelectrons. These electrons emitted from the photocathode are focused electrostatically and accelerated towards the first dynode (D_1). Each electron on striking the dynode surface causes the emission of several secondary electrons which inturn is accelerated towards the next dynode. Thus, the secondary emission from each dynode surface causes a multiplication of electrons and repeat the process until the last dynode (D_n). If δ is the average secondary electrons emitted at each dynode surface, and if there are N dynodes, the total current amplification factor between the cathode and anode is given by

$$G = \frac{I_{out}}{I_{in}} = \delta^N$$

The most important part of a photomultiplier is the photocathode which converts the incident radiation into electric current, and this determine the wavelength response characteristics. It consists of material with low work function such as compounds of silver-oxygen-cesium (Ag-O-Cs), antimony-cesium (Sb-Cs) etc. The response time of the photomultiplier is generally of the order of nanoseconds depending primarily on the dynode material used for each stage and the geometrical design of the photomultiplier structure including the gain.

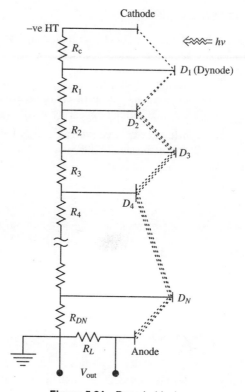

Figure 5.21 Dynode biasing.

5.11.3 Noises in PMT

Thermal Noise

Due to thermionic emission, even if there is no radiation falling on the photocathode surface, a current may by generated which often constitute the main current. This is called dark current which is of order 10^{-11} to 10^{-12} A and is a main source of noise in a PMT. The thermionic current at a temperature T, for a cathode of work functions W is given by Richardson-Dashman equation.

$$i_T = aAT^2 \exp\left(\frac{-eW}{kT}\right)$$

where A is the area of the photocathode and $a = 1.2 \times 10^6$ Am^{-2}K^{-2} for pure metal. By cooling the PMT, the dark current can be reduced.

Short Noise

Due to the discrete nature of the electronic charges, the rate of arrival of electrons at any point fluctuate and give rise to fluctuations in the current flow. The magnitude of current fluctuations for the frequency range between f and $f + \Delta f$ is given by

$$\Delta i_f = (2ie\Delta f)^{1/2}$$

where i is the sum of the dark current and the signal current.

Johnson (Nyquest) Noise

Johnson noise arises because of random nature of charge carriers due to thermal agitation which results in fluctuating voltage appearing across the conductor. The RMS value of the voltage having frequency components between f and $f + \Delta f$ across the resistor R_L at a temperature T is given by

$$\Delta V_f = (4kTR_L\Delta f)^{1/2}$$

The Johnson noise is smaller than the noise due to the dark current.

Multiplication noise

Multiplication noise arises as a result of statistical spread in the secondary electron emission about a mean value.

The presence of above noise currents set a limit on the minimum detectable signal level. If the optical signal is smaller than the noise signal, it will be difficult to detect without further processing. If we eliminate the noises all together, the average current is due to optical signal alone. The average photocurrent due to optical power is given by

$$i = \frac{e\eta P}{h\nu}$$

The minimum signal power in the presence of thermionic dark current is given by

$$W_{\min} = \frac{(2i_Te\Delta f)^{1/2}}{R_\lambda}$$

where

$$R_\lambda = \frac{i}{P}$$

is the responsivity which is the current flowing per optical power falling on the photocathode.

EXAMPLE 5.1 An argon ion laser beam with a power 1 watt and having a wavelength 514.5 nm is falling on a photodetector. Calculate the number of photons hitting the surface of the detector in every second.

Solution $\lambda = 514.5$ nm; $P = 1$ W

$$P = \frac{nh\nu}{t}$$

$$= \frac{n}{t}\frac{hc}{\lambda};$$

$$\frac{n}{t} = \frac{1 \times 514.5 \times 10^{-9}}{6.626 \times 10^{-34} \times 3 \times 10^8}$$

$$= 2.59 \times 10^{18} \text{ photons/sec.}$$

EXAMPLE 5.2 The energy required to excite electron-hole pairs in a photoconductive detector is about 0.04 eV. Calculate the cut-off wavelength at the long wavelength end.

Solution
$$\lambda_{max} = \frac{hc}{E}; \quad E = 0.04 \times 1.6 \times 10^{-19} \text{ J}$$
$$= \frac{6.626 \times 10^{-34} \times 3 \times 10^8}{0.04 \times 1.6 \times 10^{-19}}$$
$$= 31.06 \text{ μm}$$

EXAMPLE 5.3 What is the band gap of the semiconductor material in eV if it gives a maximum response at 1.06 μm?

Solution
$$E(\text{eV}) = \frac{hc}{\lambda e}; \quad \lambda = 1.06 \times 10^{-6} \text{ m}$$
$$= \frac{6.626 \times 10^{-34} \times 3 \times 10^8}{1.06 \times 10^{-6} \times 1.6 \times 10^{-19}}$$
$$= 1.17 \text{ eV}$$

EXAMPLE 5.4 Quantum efficiency of a photoconductive detector is 8%. Calculate the rate of generation of charge carriers with a light of wavelength 632.8 nm and having a power 2.5 μW.

Solution
$$G = \eta \frac{P\lambda}{hc}$$

where $P = 2.5 \times 10^{-6}$ W, $\lambda = 632.8 \times 10^{-9}$ m and $\eta = 0.08$.

$$G = \frac{0.08 \times 2.5 \times 10^{-6} \times 632.8 \times 10^{-9}}{6.626 \times 10^{-34} \times 3 \times 10^8}$$
$$= 6.367 \times 10^{11} \text{ carriers per second.}$$

EXAMPLE 5.5 The quantum efficiency of a photoconductor is 10%, the lifetime of the carrier and the average time for a carrier to cross the detector chip is 10^{-5} and 10^{-7} sec, respectively. Calculate the voltage signal when a load resistance of 100 K is used. The power of the beam is 1 μW and the wavelength is 1.06 μm.

Solution
$$i = \eta \frac{e}{h\nu} \left(\frac{\tau_0}{\tau_D} \right) P$$

where $\tau_0 = 10^{-5}$ s, $\tau_D = 10^{-7}$ s, $\eta = 0.1$, $P = 1$ μW, $\lambda = 1.06$ μm

$$i = 0.1 \times \frac{1.6 \times 10^{-19} \times 1.06 \times 10^{-6}}{6.626 \times 10^{-34} \times 3 \times 10^8} \left(\frac{10^{-5}}{10^{-7}} \right) 1 \times 10^{-6}$$
$$= 8.53 \times 10^{-6} \text{ Amp.}$$

Voltage across 100 K $= i \times R$
$$= 8.53 \times 10^{-6} \times 100 \times 10^3$$
$$= 0.853 \text{ V}$$

EXAMPLE 5.6 Show that if E_g is the band gap of a photoconductor in eV, then the cut-off wavelength is given by

$$\lambda_{max} = \frac{1.24}{E_g} \, \mu m$$

Solution
$$\lambda_{max} = \frac{hc}{E_g e}$$
$$= \frac{6.626 \times 10^{-34} \times 3 \times 10^8}{E_g 1.6 \times 10^{-19}}$$
$$= \frac{1.24}{E_g} \, \mu m$$

EXAMPLE 5.7 A photoconductor is connected in series with a resistance R_L and a bias voltage V_b is applied. If the photocurrent generated is linearly proportional to the light intensity I, show that the equivalent detector resistance varies as

$$R_d = \frac{V_b}{aI} - R_L$$

where a is a constant of proportionality.

Solution $V_b = i(R_d + R_L)$; photocurrent, $i = aI$

$$\therefore \qquad R_d = \frac{V}{aI} - R_L$$

EXAMPLE 5.8 A beam having an intensity I and frequency v is falling on a photoconductor of surface area $w \times l$ and thickness d. If η is the quantum efficiency, show that the rate of generation of charge carriers per unit volume is given by

$$G_v = \eta \frac{I}{hvd}$$

Solution
$$G = \eta \frac{p}{hv}$$
$$= \eta \frac{I \cdot area}{hv}$$
$$= \eta \frac{Iwl}{hv}$$
$$G_v = \frac{G}{Volume}$$
$$= \frac{G}{\omega l d}$$

$$\therefore \qquad G_v = \eta \frac{I}{hvd}$$

EXAMPLE 5.9 Responsivity of a Si PIN photodector is 0.5 A/W. A reverse voltage of 20 V, through a load resistance of 100 Ω is applied to it. Calculate the maximum detectable power level and sensitivity.

Solution
$$P_{max} = \frac{V_b}{\rho R_L}$$

where, $V_b = 20$ V, $\rho = 0.5$ A/W, $R_L = 100$ Ω

$$P_{max} = \frac{20}{0.5 \times 100}$$
$$= 0.4 \text{ W}$$

$$\text{Sensitivity} = \frac{V_b}{P_{max}} = \rho R_L$$
$$= 0.5 \times 100$$
$$= 50 \text{ mV/mW}$$

EXAMPLE 5.10 Assume a PIN photodiode with 100% antireflection coating and all the charge carriers contributing to the photocurrent. The absorption coefficient at 800 nm is 10^5 m^{-1}. What should be the depletion layer thickness, if the quantum efficiency is 0.7.

Solution $\eta = (1 - R) \, \delta\{1 - \exp[-\alpha(\lambda)w]\}$
$R = 0$, $\delta = 1$, $\eta = 0.7$, $\alpha = 10^5$ m^{-1}, $w = ?$

$$w = \frac{1}{\alpha} \log\left(\frac{1}{1-\eta}\right)$$
$$= \frac{1}{10^5} \log\left(\frac{1}{1-0.7}\right)$$
$$= 1.2 \times 10^{-5} \text{ m}$$

EXAMPLE 5.11 A PIN photodiode has a quantum efficiency 0.6. Calculate the responsivity at 1550 nm.

Solution
$$\rho = \eta \frac{e}{h\nu}$$
$$= \frac{0.6 \times 1.6 \times 10^{-19} \times 1550 \times 10^{-9}}{6.626 \times 10^{-34} \times 3 \times 10^8}$$
$$= 0.75 \text{ A/W}$$

EXAMPLE 5.12 In an InGaAs PIN photodiode the width of the depletion region is 5 μm and the drift velocity of the electron is 10^5 ms^{-1}. Calculate the transit time.

Solution
$$\tau_t = \frac{w}{v_d}$$
$$= \frac{5 \times 10^{-6}}{10^5} = 50 \text{ ps}$$

EXAMPLE 5.13 Calculate the time taken by electrons to diffuse through 5 μm thick p-type silicon layer in PIN photodiode. The diffusion constant is 3.4×10^{-3} m²/s.

Solution
$$\tau_{\text{diff}} = \frac{d^2}{2D_c}$$
$$= \frac{(5 \times 10^{-6})^2}{2 \times 3.4 \times 10^{-3}}$$
$$= 3.68 \times 10^{-9} \text{ s}$$

EXAMPLE 5.14 A silicon PIN photodetector with radius of 500 μm, depletion width 20 μm and having permittivity 10.5×10^{-13} F/cm. Calculate the rise time and corresponding bandwidth. The load resistance 1 K.

Solution $A = \pi r^2$, $r = 500$ μm, $\varepsilon = 10.5 \times 10^{-13}$ F/cm, $w = 20$ μm and $R_L = 1$ K.
$$c_d = \frac{\varepsilon A}{w}$$
$$= 4.1 \text{ pF}$$
$$t_r = 2.19 R_L \frac{\varepsilon A}{w} = 9 \text{ ns}$$
$$\Delta f = \frac{1}{2\pi R_L c_d}$$
$$= 38.8 \text{ MHz}$$

EXAMPLE 5.15 A phototransistor with $\alpha = 0.98$ and $I_{CBO} = 0.1$ μA having an area of 0.4 cm² and quantum efficiency 90% is irradiated with a beam of 1 W/m² and wavelength 1500 nm. Calculate the emitter current.

Solution $\eta = 0.9$, $A = 0.4 \times 10^{-4}$ m², $I = 1$ Wm⁻², $\lambda = 1500 \times 10^{-9}$ m and $I_{CBO} \cong 0$
$h_{fe} = \alpha/(1 - \alpha)$; $\alpha = 0.98$
$$\therefore \quad h_{fe} = 49$$
$$I_\lambda = \frac{\eta A e I \lambda}{hc}$$
$$= 4.35 \times 10^{-5} \text{ Amp.}$$
$$\therefore \quad I_E = I_\lambda (h_{fe} + 1) = 2.17 \text{ mA}$$

EXAMPLE 5.16 Calculate the responsivity of a PMT having a quantum efficiency of 80% at 600 nm.

Solution
$$R_\lambda = \frac{\eta e \lambda}{hc}$$
$$= \frac{0.8 \times 1.6 \times 10^{-19} \times 600 \times 10^{-9}}{6.626 \times 10^{-34} \times 3 \times 10^8}$$
$$= 0.386 \text{ A/W}.$$

EXAMPLE 5.17 Estimate the dark current, if the work function of the cathode material is 1.1 eV and area 4 cm² at 27°C.
($a = 1.2 \times 10^6$ Am^{-2}K^{-2}, $k = 1.381 \times 10^{-23}$ JK^{-1})

Solution $i_T = aAT^2 \exp\left(\dfrac{-eW}{kT}\right)$ $T = 300$ K

$= 15.35$ pA

EXAMPLE 5.18 Calculate the gain of a PMT if there are nine dynodes in it and the secondary electron generation factor is 5.

Solution $G = \delta^N = 5^9 = 1.95 \times 10^6$

EXAMPLE 5.19 The bandwidth of a PMT is 200 KHz and the responsivity is 0.15 AW^{-1}. If the dark current is 0.15×10^{-13} A. Calculate the minimum detectable power level.

Solution $W_{min} = \dfrac{(2i_T e \Delta f)^{1/2}}{R_\lambda}$

$= \dfrac{(2 \times 0.15 \times 10^{-13} \times 1.6 \times 10^{-19} \times 200 \times 10^3)^{1/2}}{0.15}$

$= 2 \times 10^{-13}$ W

EXAMPLE 5.20 Calculate the maximum wavelength that has sufficient energy to eject electron from a photocathode of work function of 2.5 eV.

Solution $\lambda_{max} = \dfrac{hc}{Ee}$

$= \dfrac{6.626 \times 10^{-34} \times 3 \times 10^8}{2.5 \times 1.6 \times 10^{-19}}$

$= 497$ nm

REVIEW QUESTIONS

1. What are photodetectors? What is the general working principle of a photodetector?
2. What are the basic processes that involved in a photon detectors?
3. Mention the performance requirement of a photodetector.
4. Explain the principle of photoconductive detectors.
5. Explain the structure of a CdS photoconductor.
6. Mention some photoconductive semiconductor materials.
7. Explain the working of a photodiode.
8. Draw V-I characteristics of a photodiode.
9. Explain the design of a photodiode.
10. Describe the equivalent circuit of a photodiode and obtain the expression for photocurrent.

11. Explain the PIN photodiode and obtain the expression for photocurrent.
12. Define responsivity of a photodiode.
13. Explain response time of a PIN photodiode.
14. Explain the working of an APD.
15. Explain the design of an APD.
16. Explain the working and the structure of
    (a) Modulated barrier photodiode
    (b) Schottky barrier (MSM) photodiode
    (c) Microcavity photodiode
17. Explain the working principle of a phototransistor.
18. Explain photoelectric effect.
19. Explain the design and working of a PMT.
20. Write short notes on
    Thermal noise
    Short noise
    Johnson noise
    Multiplication noise

6

Solar Cells

6.1
INTRODUCTION

Energy radiated from sun is in the form of electromagnetic radiation which falls in earth mainly in the spectral region UV to IR (0.2 µm to 3 µm). The other radiations including radio waves are also present but with a less intensity. Part of this radiation reaching to the earth from the sun is scattered or absorbed by the earth's atmosphere. Depending on the angle of incidence, the intensity of radiation varies between 500 to 1000 W/m^2. Conversion of optical energy into electrical energy is known as photovoltaic effect. Solar cell is such a device in which solar energy is directly converted into electric energy. Solar cell is an important photovoltaic device for obtaining energy from the sun. It can provide nearly permanent power at low operating cost which is almost free of pollution.

Primary requirement for a material to be used as a solar cell is a bandgap matching material for the solar spectrum with high mobilities and lifetime for charge carriers. Solar cells using many semiconductors with various configurations employing single crystal, polycrystal and amorphous thin film structure have been used.

More than 95% of the solar cells in production are silicon based. A single crystal junction solar cell is a *p-n* junction. When light falls on an unbiased solar cell, electron-hole pairs are created, that diffuse towards the junction and an electrical power develops across the junction as in the battery and this power is delivered to an external load.

6.2
PRINCIPLE OF OPERATION

The photovoltaic energy conversion process may be expressed by the equivalent circuit shown in Figure 6.1. An ideal diode is connected in parallel with a constant current source which represents the photovoltaic energy generated, and with a load resistor.

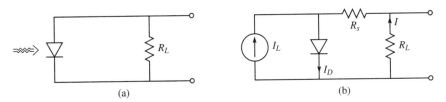

Figure 6.1 (a) Photovoltaic effect, (b) Equivalent circuit.

Consider a conventional solar cell exposed to solar radiation. Let the photon energy be greater than the bandgap energy of the semiconductor material. Part of this energy equivalent to the bandgap energy is contributed to the cell output and the remaining is converted into heat. In the equivalent circuit a current source is connected in parallel with the junction. The photogenerated current I_L results from the excitation of excess carriers. The IV characteristics of the solar cell is represented by

$$I = I_D - I_L = I_s \left[\exp\left(\frac{eV}{kT}\right) - 1 \right] - I_L$$

where, I_s is the diode saturation current. In the dark, drift of thermally generated minority carriers across the junction constitutes the reverse saturation current. At zero bias this is balanced by a small flow of majority carriers in the opposite direction resulting to zero net current. If the diffusion lengths of holes and electrons are L_h and L_e, their lifetimes τ_h and τ_e, with the concentration per unit volume is given by p_0 and n_0, respectively. Then the reverse saturation current is given by

$$I_s = eA \left(\frac{L_h}{\tau_h} p_0 + \frac{L_e}{\tau_e} n_0 \right)$$

Here A is the area of the diode surface. When the junction is illuminated by photon energy $h\nu \geq E_g$ additional electron-hole pairs are created with a generation rate G per unit volume. Then the number of holes created per second is $AL_h G$ and the number of electrons created per second is $AL_e G$. The total photogenerated current I_L due to the drift of these carriers across the junction is given by

$$I_L = eAG (L_h + L_e)$$

This current is opposite to the main current. Thus

$$I = eA \left(\frac{L_h}{\tau_h} p_0 + \frac{L_e}{\tau_e} n_0 \right) \left[\exp\left(\frac{eV}{kT}\right) - 1 \right] - eAG(L_h + L_e)$$

The IV characteristics of the solar cell is shown in the Figure 6.2(a). It passes through the fourth quadrant as shown in Figure 6.2(b).

Consider the short circuited diode condition for $V = 0$, the short circuited current

$$I_{SC} = -eAG (L_h + L_e)$$

For the open circuit diode $I = 0$, so that with $V = V_{oc}$

$$eA \left(\frac{L_h}{\tau_h} p_0 + \frac{L_e}{\tau_e} n_0 \right) \left[\exp\left(\frac{eV_{oc}}{kT}\right) - 1 \right] - eAG(L_h + L_e) = 0$$

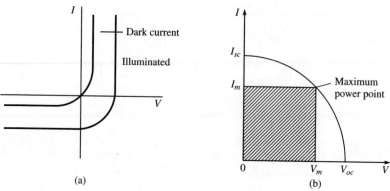

Figure 6.2 (a) *IV* characteristic of a solar cell, (b) *IV* characteristics of an illuminated solar cell in the fourth quadrant.

or

$$\exp\left(\frac{eV_{oc}}{kT}\right) - 1 = \frac{I_L}{I_s}$$

$$\frac{eV_{oc}}{kT} = \log\left(\frac{I_L}{I_s} + 1\right)$$

$$V_{oc} = \frac{kT}{e} \log\left(\frac{I_L}{I_s} + 1\right)$$

For $I_L \gg I_s$

$$V_{oc} \approx \frac{kT}{e} \log\left(\frac{I_L}{I_s}\right)$$

Under illumination the output electrical power

$$P = IV$$

$$= \left\{ I_s \left[\exp\left(\frac{eV}{kT}\right) - 1 \right] - I_L \right\} V$$

$$= I_s V \left[\exp\left(\frac{eV}{kT}\right) - 1 \right] - I_L V$$

The condition for maximum power is $\left.\dfrac{dP}{dV}\right|_{V=V_{max}} = 0$

$$I_s \left[\exp\left(\frac{eV_m}{kT}\right) - 1 \right] + I_s \frac{eV_m}{kT} \exp\left(\frac{eV_m}{kT}\right) = I_L$$

$$\exp\left(\frac{eV_m}{kT}\right)\left(1 + \frac{eV_m}{kT}\right) = 1 + \frac{I_L}{I_s}$$

$$\exp\frac{eV_m}{kT} = \frac{1 + (I_L/I_s)}{1 + (eV_m/kT)}$$

or
$$V_m = \frac{kT}{e} \log\left[\frac{1+(I_L/I_s)}{1+(eV_m/kT)}\right]$$
$$= \frac{kT}{e} \log\left(1 + \frac{I_L}{I_s}\right) - \frac{kT}{e} \log\left(1 + \frac{eV_m}{kT}\right)$$
$$= V_{oc} - \frac{kT}{e} \log\left(1 + \frac{eV_m}{kT}\right)$$

The current corresponding to the maximum value of power is given by

$$I_m = I_s\left[\exp\left(\frac{eV_m}{kT}\right) - 1\right] - I_L$$
$$= I_s\left[\exp\left(\frac{eV_m}{kT}\right) - 1\right] - I_s\left[\exp\left(\frac{eV_m}{kT}\right) - 1\right] - I_s \frac{eV_m}{kT} \exp\frac{eV_m}{kT}$$
$$= -I_s \frac{eV_m}{kT} \exp\frac{eV_m}{kT}$$

The magnitude of I_m is thus can be expressed as

$$I_m = I_s \frac{eV_m}{kT} \exp\frac{eV_m}{kT}$$
$$= I_s \frac{eV_m}{kT} \frac{1+(I_L/I_s)}{1+(eV_m/kT)}$$
$$= I_s\left(\frac{I_L + I_s}{I_s}\right) \frac{eV_m}{kT} \frac{1}{(eV_m/kT)[1+(kT/eV_m)]}$$
$$\approx I_L\left(1 + \frac{kT}{eV_m}\right)^{-1}$$
$$\approx I_L\left(1 - \frac{kT}{eV_m}\right)$$

The maximum power is given by

$$P_m = I_m V_m$$
$$= I_L\left(1 - \frac{kT}{eV_m}\right) V_m$$
$$= I_L\left(V_m - \frac{kT}{e}\right)$$
$$= I_L\left[V_{oc} - \frac{kT}{e} \log\left(1 + \frac{eV_m}{kT}\right) - \frac{kT}{e}\right]$$
$$= I_L\left(\frac{E_m}{e}\right)$$

where

$$E_m = e\left[V_{oc} - \frac{kT}{e}\log\left(1 + \frac{eV_m}{kT}\right) - \frac{kT}{e}\right]$$

corresponds to the energy per photon delivered to the load at the maximum power point.

The ideal conversion efficiency of the solar cell η is defined as the ratio of the maximum power output to the incident power

$$\eta = \frac{P_{max}}{P_{in}} = \frac{V_m I_m}{P_{in}}$$

$$= \frac{V_m^2 I_s (e/kT) \exp(eV_m/kT)}{P_{in}}$$

Another useful quantity in specifying the solar cell characteristics is the fill factor which is a measure of power extraction efficiency and is defined as

Fill factor, $$F_f = \frac{I_m V_m}{I_{sc} V_{oc}}$$

or $$P_m = F_f I_{sc} V_{oc}$$

To maximize output power both I_{sc} and V_{oc} must be made as large as possible. For the most solar cells, the fill factor is about 0.7 which is an important figure of merit in its design.

6.3
SPECTRAL RESPONSE

The spectral response of a solar cell is the short circuit current as a function of wavelength of the incident light. When a monochromatic light of wavelength λ is incident on the front surface of a solar cell, the rate of generation of electron hole pairs is given by

$$G(\lambda, x) = \alpha(\lambda) F(\lambda) [1 - R(\lambda)] \exp[-\alpha(\lambda)x]$$

where $\alpha(\lambda)$ is the absorption coefficient, $F(\lambda)$ is a distribution function which gives the number of incident photons per area per second per unit bandwidth and $R(\lambda)$ is the fraction of these photons reflected from the surface. The photocurrent at a wavelength λ is the resultant of the following:

1. Photocurrent due to electrons collected at the depletion edge—$J_n(\lambda)$
2. Photocurrent due to holes collected at the depletion edge—$J_p(\lambda)$
3. The photocurrent due to the carriers within the depletion region—$J_{dr}(\lambda)$

Thus $$J(\lambda) = J_n(\lambda) + J_p(\lambda) + J_{dr}(\lambda)$$

Thus total current density is in the whole wavelength range

$$J = e \int_0^{\lambda_m} F(\lambda)[1 - R(\lambda)] s(\lambda)\, d\lambda$$

where λ_m is the largest wavelength corresponding to the semiconductor bandgap and $s(\lambda)$ is the internal spectral response function given by

$$s(\lambda) = \frac{J(\lambda)}{eF(\lambda)[1-R(\lambda)]}$$

The ideal response curve of a solar cell with energy gap E_g is a step function. It is equal to zero for $v < E_g/h$ and unity for $v \geq E_g/h$. In reality the spectral response depends on the device parameters particularly the absorption coefficient so that an upper and lower cut-off is observed as in Figure 6.3.

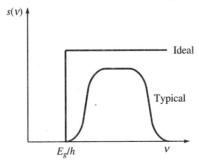

Figure 6.3 Spectral response of a solar cell.

6.4
HOMOJUNCTION SOLAR CELLS

Figure 6.4 shows the schematic representation of a silicon p-n junction solar cell. It consists of a *p-n* junction with a front ohmic contact stripe, a back contact that covers the entire back surface and an anti-reflection coating on the front surface. By using light concentration system the cell efficiency can be improved and also reduce the overall cost.

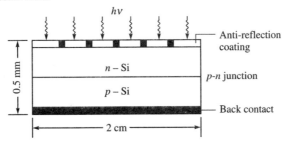

Figure 6.4 Homojunction solar cell design.

6.5
HETROJUNCTION SOLAR CELLS

A junction formed between two or more semiconductors with different bandgap is called a hetrojunction. The *IV* characteristics of a hetrojunction solar cell is very similar to that of a homojunction cell. Basically, it consists of two semiconductor layers with band gaps E_{g1} and E_{g2}. Light with energy less than E_{g1} but greater than E_{g2} will pass through the first semiconductor which acts as a window. Light with energy greater than E_{g1} will be absorbed by the first semiconductor and the carriers generated within the diffusion length from the junction or within the depletion region will be collected similar to a *p-n* homojunction. The long wavelength cut-off is given by E_{g2} and the short wavelength response depends on the bandgap and the thickness of the first semiconductor. Some advantages of the hetrojunctions solar cells over conventional junction cells are

1. Enhance the short wavelength spectral response. High energy photons can be made to absorb inside the depletion region of the second semiconductor by making the band gap E_{g1} large enough.
2. Series resistance of the cell is greatly reduced if the higher band gap material is doped heavily without affecting its transmission characteristics.
3. The radiation tolerance of the solar cell is improved if a thick high band gap window layer is used.

6.6

AlGaAs/GaAs HETROJUNCTION SOLAR CELLS

Schematic of a *p*-GaAlAs/*p*-GaAs/*n*-GaAs hetrostructure solar cell is shown in Figure 6.5. The AlGaAs acts as a transparent window admitting photons of energy less than the band gap of AlGaAs (E_{g1}). The AlGaAs layer band gap is larger than that of GaAs (E_{g2}). Those photons with energies between E_{g1} and E_{g2} will create carriers and the absorption is high in the GaAs. The carriers are generated in the depletion region or close to it. So that the collection is high.

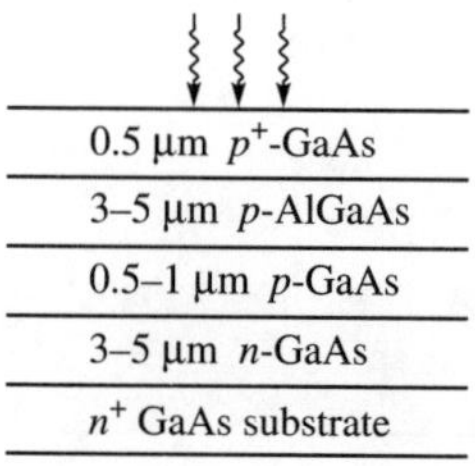

0.5 µm p^+-GaAs
3–5 µm *p*-AlGaAs
0.5–1 µm *p*-GaAs
3–5 µm *n*-GaAs
n^+ GaAs substrate

Figure 6.5 Schematic of a typical hetrostructure solar cell.

6.7

THIN FILM SOLAR CELLS

In thin film solar cells, the active semiconductor layers are polycrystalline or disordered films that have been deposited or formed on substrates such as glass, plastic, ceramic, metal, graphite or metallurgical silicon. In a typical example, thin film CdS solar cells are fabricated using a substrate of electroformed copper, coated with 0.5 µm of zinc. A layer of CdS about 20 µm thick is evaporated on the heated substrated at 220°C. Reacting the CdS film in a cuprous in solution form a Cu_2S layer of 1000 Å. A transparent grid contact is deposited on Cu_2S and an anti-reflection layer is applied over Cu_2S.

Under front illumination, most light is absorbed in Cu_2S. The spectral response and photocurrent are limited by high surface recombination velocity, short diffusion length and high recombination. These solar cells have conversion efficiency near 10% and can be increased up to 14% by substituting zinc for 15 to 25% of the cadmium.

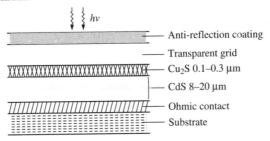

Figure 6.6 Thin film solar cell.

Thin film of Si, GaAs, InP, CdTe, $CuInSe_2$/CdS hetrojunction etc. can be deposited by various methods such as vapour growths, evaporation electroplating, etc. The main advantages of thin film solar cells is that their low cost due to low cost processing and the use of relatively low cost materials. The main disadvantages are low efficiency and long term instability.

6.8
SCHOTTKY–BARRIER SOLAR CELLS

It is a metal—n-semiconductor (or a metal-p semiconductor). The metal film is deposited on a n-semiconductor and a transparent film through which most of the light can pass. The metal film must be thin enough to allow a substantial amount of light to reach the semiconductor. The IV characteristics of a Schottky barrier under illumination can be represented by

$$I = I_s \left[\exp\left(\frac{eV}{kT}\right) - 1 \right] - I_L$$

where

$$I_s = Aa^* T^2 \exp\left(\frac{-e\phi_B}{kT}\right)$$

where A is the area a^* is the effective Richardson constant and $e\phi_B$ is the barrier height. The maximum efficiency is about 25%. To achieve high barrier height on semiconductor one usually uses metal with high work function for n-type and low work function for p-type semiconductors. The advantages of Schottky barrier solar cells are as follows:

1. Low temperature processing is involved because no high temperature diffusion is required.
2. Adaptability to polycrystalline material
3. High radiation resistance
4. High current output and good spectral response.

6.9
CASCADE HETROJUNCTION SOLAR CELLS

It has been proposed that the efficiency can be greatly enhanced by utilizing multi-step tunnel process in the hetrojunction. The cascade hetrojunction solar cell consists of a wide gap (1.59 eV) and a narrow gap (0.95 eV) cells joined by a hetrojunction tunnel diode formed as an integral part of a monolithic structure. The design also includes a heteroface window layer to minimize surface recombination losses. Light passes through the first cell without being absorbed. It then goes through the ultra thin tunnel diode and collected by the narrow gap cell. The efficiency up to 30% can be achieved at room temperature with this structure.

6.10
MATERIAL REQUIREMENTS

The requirement of a material to be used for solar cell fabrication is that it should be a band gap matching material to the solar spectrum and also should have high mobilities and lifetime of charge carriers. These conditions are met by Si, GaAs, CdSe, CdTe, etc., and many II–VI and III–V compounds. Solar cells are also made of polycrystalline and amorphous semiconductors, though their efficiency is less than that of single crystal ones. They are low cost and are mainly used for large scale applications.

Cost of a cell depends roughly on the relative abundance of the material used and their melting point. In general, for terrestrial applications, the cost reduction is an important factor where cheaper materials with less expensive manufacturing processes are used, whereas for space applications, cost not considered as an important factor.

6.11
TEMPERATURE AND RADIATION EFFECTS ON SOLAR CELL PERFORMANCE

As the temperature increases, the diffusion length increases, which result into an increase in I_L. However, since the saturation current increases exponentially with the temperature, V_{oc} decreases rapidly. The overall effect is a reduction of the cell efficiency with the increase in temperature.

For space applications, it is important to consider radiation effects. The high energy radiation in the outer space can produce deflects in semiconductors that cause reduction in solar cell power output. To improve the radiation tolerance certain elements such as Lithium has been incorporated into the solar cells which reduces the degradation of lifetime.

6.12
OPTICAL CONCENTRATION

Optical concentration of solar radiation offers an attractive and flexible approach to reduce the cell cost by substituting a concentrator area for the cell using mirrors, lenses, etc. It also offers advantages as follows:

1. Increased cell efficiency
2. Hybrid system yielding both electrical and thermal outputs.
3. Reduced cell temperature coefficient
4. Expensive cell material can be replaced by adopting less expensive concentrator materials which minimizes the overall system gain.

In a typical concentration module shown in Figure 6.7 sunlight is reflected by a primary mirror to a secondary mirror which is then focused onto the solar cell. The device is mounted on a water-cooled block which considerably improves the device performance.

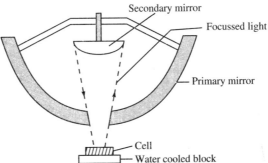

Figure 6.7 Solar cell optical concentration.

6.13
DEVICE FABRICATIONS

Many solar cell configurations have been proposed for achieving high conversion efficiency some of them are briefly described below.

Back Surface Field (BSF) Solar Cell

BSF cells have much larger output voltage than the conventional cells. The front surface of a n^+-p-p^+ BSF junction solar cell is made in the normal way but back of the cell instead of containing just a metallic ohmic contact.

Violet Cell

The violet cell has a much higher response in the violet region ($hv > 2.75$ eV). They are fabricated with reduced surface doping concentration and smaller junction depth.

Textured Cell

They have pyramidal surface with anti-reflection coating. Light incident on the side of a pyramid will be reflected to another, instead of being lost. This enhances both the short circuit current and open circuit voltage which in turn gives cell efficiency over 15%.

V-groove Multijunction Solar Cells

This cell consists of many p^+-n-n^+ or p^+-p-n^+ diode elements connected in series. The effective optical thickness is much higher which enhances the collection efficiency up to 93%. The conversion efficiency is about 20%.

Tandom Junction Solar Cells (TJC)

The TJC combines the concepts of both BSF and textured cell which is expected to have cell efficiency above 20%.

Vertical Junction Solar Cell

They have both the junction and the metallization perpendicular to the cell surface. They have efficiency above 15%.

EXAMPLE 6.1 A silicon solar cell with a diode area 10^{-4} cm^2 when irradiated, the electron-hole pair generation rate was 10^{22} s^{-1}cm^{-3}. The diffusion length of electrons and holes are 5×10^{-4} and 4×10^{-4} cm, respectively. Calculate the photogenerated current.

Solution $\qquad I_L = eAG(L_e + L_h)$

where $L_e = 5 \times 10^{-4}$ cm, $L_h = 4 \times 10^{-4}$ cm, $G = 10^{22}$ s^{-1}cm^{-3} and $A = 10^{-4}$ cm^2.

$$I_L = 1.6 \times 10^{-19} \times 10^{-4} \times 10^{22}(5 \times 10^{-4} + 4 \times 10^{-4})$$
$$= 0.144 \text{ mA}$$

EXAMPLE 6.2 In a Si solar cell at 300 K the photogenerated current and the reverse saturation current are found to be 25 mA and 40 PA, respectively. Calculate the open circuit voltage ($k = 1.38 \times 10^{-23}$ SI units)

Solution $\qquad V_{oc} = \dfrac{kT}{e} \log_e \left(\dfrac{I_L}{I_s} + 1 \right)$

$$= \dfrac{1.38 \times 10^{-23} \times 300}{1.6 \times 10^{-19}} \log_e \left(\dfrac{25 \times 10^{-3}}{40 \times 10^{-12}} + 1 \right)$$

$$= 0.524 \text{ V}$$

EXAMPLE 6.3 For a silicon solar cell with fill factor 0.7 and the short circuit current 25 mA and the open circuit voltage 0.5 V. Calculate the maximum output power of the cell.

Solution
$$P_m = F_i I_{sc} V_{oc}$$
$$= 0.7 \times 25 \times 10^{-3} \times 0.5$$
$$= 8.75 \times 10^{-3} \text{ W}$$

EXAMPLE 6.4 A silicon solar cell with area 5×10^{-4} cm^2 when exposed to light produces 10^{23} cm^{-3}s^{-1} electron-hole pairs. The electron and the hole diffusion lengths are assumed to be 10 μm and the dark reverse saturation current to 100 pA. Calculate the photocurrent and also the open circuit voltage (T = 300 K).

Solution
$$I_L = eAG(L_e + L_h)$$
$$= 1.6 \times 10^{-19} \times 5 \times 10^{-4} \times 10^{23} \times 2 \times 10 \times 10^{-4}$$
$$= 16 \text{ mA}$$

$$V_{oc} = \frac{kT}{e} \log_e \left(\frac{I_L}{I_s}\right)$$
$$= \frac{1.38 \times 10^{-23} \times 300}{1.6 \times 10^{-19}} \log_e \left(\frac{16 \times 10^{-3}}{100 \times 10^{-12}}\right)$$
$$= 0.49 \text{ V}$$

REVIEW QUESTIONS

1. What is photovoltaic effect?
2. Explain with necessary theory the principle of solar cell, and obtain the expression for V_{oc}, V_m, I_m.
3. Define the fill factor of a solar cell.
4. Explain the spectral response of a solar cell.
5. Write the short notes on:
   (i) Homojunction solar cell
   (ii) Hetrojunction solar cell
   (iii) Thin film solar cells
   (iv) Schottky barrier solar cells
   (v) Cascade hetrojunction solar cells
6. What is the material requirements for a solar cell?
7. Explain the temperature and radiation effects on solar cell performance.
8. Explain the optical concentration of radiation in solar cells.
9. Explain the different types of solar cells.

7

Fibre Optics

7.1
OPTICAL FIBRE

An optical fibre is a transmission line (waveguide) that can guide optical radiation from one place to another. It consists of a cylindrical transparent material of refractive index n_1 in the central region called as core, which is then surrounded by an annular shaped outer region of slightly lower refractive index n_2, called as cladding. Silica glass or multicomponent glasses are usually used as dielectric materials.

To withstand chemical and mechanical strains normally a plastic primary coating and then a nylon coating are provided to the fibre as protective covering. In a typical fibre the core diameter is around a few tens of micrometer, that of cladding in the range of 100–200 μm and the thickness of the protective covering is about 3–50 μm (See Figure 7.1).

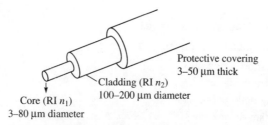

Protective covering
3–50 μm thick
Cladding (RI n_2)
100–200 μm diameter
Core (RI n_1)
3–80 μm diameter

Figure 7.1 Typical optical fibre showing core, cladding and protective covering.

An optical fibre that can transmit many optical modes is known as multimode fibre and that capable of transmitting only one mode is known as single mode fibre. Single mode fibres have very small core diameter (~2 to 8 μm) and that of the multimode fibre is of the order of 50 μm.

Depending on the usage and the requirements following types of optical fibres are used:

(i) Passive fibre: It guides light incident on it from an external source.

(ii) Active fibre: It emits light as well as guides part of it.

(iii) **Luminescent fibre:** These fibres are capable of emitting luminescent radiation when excited by UV, X-ray or high energy radiation.
(iv) **Lasing fibre:** These fibres are small diameter fibres where LASER action takes place.
(v) **Glass coated glass fibre:** It consists of glass core region with a coating of low refractive index glass.
(vi) **Conical fibre:** Conical fibre is used for light condensing and in align assemblies for magnification and demagnification purposes.
(vii) **Multiple fibre:** It consists of a multitude of smaller diameter fibres.

The refractive index of the core and cladding is one of the important factor that decides the properties of the optical fibre. The size of the core and cladding also effect their characteristics. An optical fibre cable may contain several such fibres.

7.2

TOTAL INTERNAL REFLECTION

Consider a plane interface between two media of refractive indices n_1 and n_2. When a light ray is incident on the interface it splits into two beams. The part of the incident ray reflected back into the first medium is called the reflected ray and that enter the second medium is called the refracted ray. The angles that are made by these rays with the line perpendicular to the boundary are called the angle of incidence (ϕ_1), the angle of reflection (ϕ_1) and the angle of refraction (ϕ_2) respectively. This is shown in Figure 7.2.

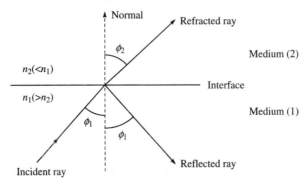

Figure 7.2 Reflection and refraction of light at a boundary separating two media.

Light when passes through different media its velocity will change. If c is the velocity of light in air and v is the velocity of light in the medium, then the ratio

$$n = \frac{c}{v}$$

is called the refractive index of the medium. Further, when light passes through media of different refractive indices, its direction will also change. This change in the direction of light (bending of light) at the boundary is called the refraction of light and is due to the difference in the speed of light in two media that have different refractive indices. The relation between the angle of incidence and the refraction is given by Snell's law

$$n_1 \sin \phi_1 = n_2 \sin \phi_2$$

Consider that the light travels towards the boundary from an optically denser region i.e., $n_1 > n_2$. Then the light in the second medium (optically rarer medium) bends away from the normal. As the angle of incidence ϕ_1 increases, the angle of refraction ϕ_2 also increases and approaches $\pi/2$. Thus, for an angle of incidence $\phi_1 = \theta_c$ is such that

$$\theta_c = \sin^{-1}\left(\frac{n_2}{n_1}\right)$$

the refracted angle is $\phi_2 = \pi/2$, i.e. the refracted ray travels along the boundary. The angle θ_c is known as critical angle of incidence. For angle of incidence greater than the critical angle, there is no refracted ray, all the light reflected back in the incident medium and the phenomenon is referred to as total internal reflection. Figure 7.3 explains the total internal reflection in an optical medium.

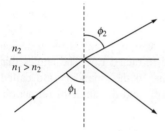

(a) Reflection, refraction and refraction at an arbitrary angle ϕ_1
$\phi_1 < \theta_c$

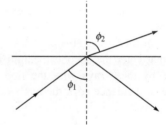

(b) The angle of incidence ϕ_1 is increased.
$\phi_1 < \theta_c$

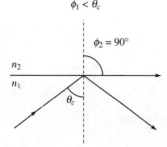

(c) At the critical angle of incidence $\phi_1 = \theta_c$.

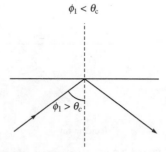

(d) The angle of incidence greater than $\phi_1 > \theta_c$.

Figure 7.3 Total internal reflection.

7.3

TYPES OF OPTICAL FIBRES

Depending on the refractive index profile of the core and cladding materials there are many types of optical fibres. A graded index fibre, has a parabolic refractive index profile, while in a step index fibre there is a sharp change in refractive index profile at the core cladding region. Other fibres with modified refractive index profile structure such as triangular profile, W profile, depressed profile, segmented profile, annular step index profile, annular graded index profile (Figure 7.4) is also under investigation. Of these step index and graded index profile fibres are commonly used.

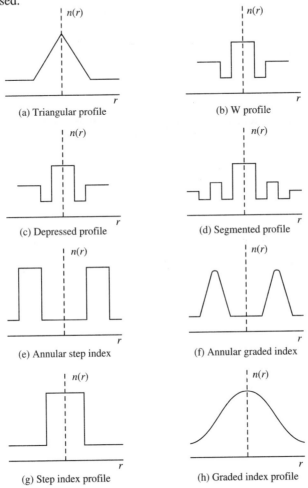

Figure 7.4 Refractive index profile of optical fibre.

7.3.1 Step Index Fibre

Refractive index of the core is uniform throughout, which is slightly higher than that in the cladding. The refractive index at the core cladding boundary changes abruptly or form a step as shown in Figure 7.5.

The corresponding refractive index distribution in the radial direction can be expressed as

$$n(r) = n_1 \quad \text{core}; \quad 0 \leq r < a$$
$$= n_2 \quad \text{cladding}; \quad r > a$$

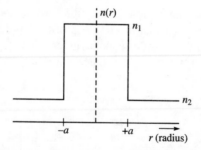

Figure 7.5 Refractive index profile of a step index fibre.

where, a is the radius of the core and r represents the cylindrical radial co-ordinate.

The difference between n_1 and n_2 is very small. The fractional change in the index of refraction is defined as

$$\Delta = \frac{n_1^2 - n_2^2}{2n_1^2} = \frac{(n_1 + n_2)(n_1 - n_2)}{2n_1^2}$$

$$\approx \frac{n_1 - n_2}{n_1}$$

Thus, the refractive index of the cladding is related to the refractive index of the core by

$$n_2 = n_1(1 - \Delta)$$

Typical values of Δ can vary between 0.001 and 0.02 (i.e. 0.1% to 2%) with a core of refractive index is 1.47.

7.3.2 Graded Index Fibre

In a graded index fibre, the refractive index decreases gradually with increasing radial distance from the centre of the fibre such that

$$n(r) = n_1 \sqrt{1 - 2\Delta \left(\frac{r}{q}\right)^q} \; ; \quad 0 \leq r \leq |a|$$

$$= n_1 \sqrt{1 - 2\Delta} = n_2 \; ; \quad r > |a|$$

where n_1 is the refractive index at the centre. The fractional change in the index of refraction is given by

$$\Delta = \frac{n^2(0) - n_2^2}{2n^2(0)} = \frac{[n(0) + n_2][n(0) - n_2]}{2n^2(0)}$$

$$= \frac{n_1 - n_2}{n_1}$$

which is same as in the case of step index fibre.

The quantity q is called the profile parameter or the gradient of the graded index fibre. $q = 2$ corresponds to a parabolic profile $q = 1$ is a triangular profile and

$q = \infty$ is a step index profile. The refractive index variation in a graded index fibre is shown in Figure 7.6.

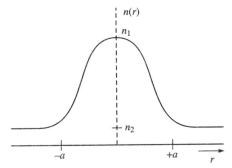

Figure 7.6 Refractive index profile of a graded index fibre.

7.4
RAYS IN AN OPTICAL FIBRE

There are two types of rays that can propagate in an optical fibre. They are meridional rays and skew rays. Meridional rays pass through the axis of the fibre. They are confined to the meridional planes, which are the planes that contain the axis of symmetry of the fibre. For example, when a ray is launched in the xz plane on the z-axis, meridional rays confined in that plane and intersect the z-axis. Meridional rays are further classified as bound rays and unbound rays. As the ray propagates along the fibre axis, the bound rays are trapped in the core, where as the unbound rays are reflected from outside of the fibre core.

Skew rays propagate in helical path along the fibre. They are not confined in a single path. Since meridional rays lie in a single plane, ray tracing of these rays is simple as compared with skew rays.

7.5
NUMERICAL APERTURE

Consider an optical fibre with index of refractions at the core n_1 and at the cladding n_2. Let a ray be launched into the fibre from a medium of refractive index n, at an angle θ_m such that the ray suffers total internal reflection at the core-cladding interface of the fibre. By applying Snell's law to the entering ray at the medium-fibre face.

$$n \sin \theta_m = n_1 \sin\left(\frac{\pi}{2} - \theta_c\right)$$
$$= n_1 \cos \theta_c$$
$$= n_1 (1 - \sin^2 \theta_c)^{1/2}$$
$$= n_1 \left(1 - \frac{n_2^2}{n_1^2}\right)^{1/2}$$

$$n \sin \theta_m = \sqrt{n_1^2 - n_2^2} \equiv NA \qquad (7.1)$$

A ray having entrance angle less than α_m will be totally internally reflected at the core-cladding interface of the fibre. $n \sin \theta_m$ is called the Numerical Aperture (NA) of the fibre. NA of a fibre is a measure of its light gathering power from a source. It is a measure of the maximum acceptance angle of the fibre.

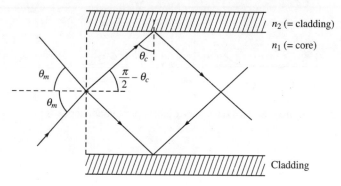

Figure 7.7 Numerical aperture.

If the entrance medium is air then $n = 1$, so that, the numerical aperture

$$NA = \sin \theta_m = (n_1^2 - n_2^2)^{1/2}$$

$$\therefore \quad \theta m_a = \sin^{-1} = (n_1^2 - n_2^2)^{1/2}$$

Equation implies that to change NA we have to change either n_1 or n_2. But

$$NA = (n_1^2 - n_2^2)^{1/2}$$
$$= [(n_1 + n_2)(n_1 - n_2)]^{1/2}$$
$$= [2n_1(n_1 - n_2)]^{1/2}$$
$$= \sqrt{n_1^2 \, 2\left(\frac{n_1 - n_2}{n}\right)}$$
$$= n_1 \sqrt{2\Delta}$$

This shows that the average and the relative difference of the index of refractions of the core and cladding are the important terms that determine the numerical aperture of the fibre. Above relations are applicable to meridional rays in an ordinary fibre only. NA is an important parameter that describe the source to optical power coupling efficiency, modes excitation, connector efficiency, splice loss, etc.

7.6
TYPE OF MODES

Propagation of electromagnetic waves in an optical fibre can be described by Maxwell's equations. The solutions of the wave equation applying appropriate boundary conditions give the electromagnetic waves that can propagate within the waveguide structure. These electromagnetic waves are known as modes. In principle, though a large number of waves can exist, only certain mode can support and can propagate within the fibre. The simplest ones are the Transverse Electric (TE) modes (electric field vector transverse to the direction of propagation are called TE modes) and the Transverse Magnetic (TM) modes (The magnetic field vector transverse to the direction of propagation are called TM modes). Combinations of TE and TM modes called hybrid modes and Linearly Polarized (LP) modes can also propagate in the fibre.

Different types of modes in an optical fibre are as follows:

7.6.1 Guided Modes

The guided modes are trapped in the fibre core. They propagate in the fibre by a large number of total internal reflection over the length of the fibre and reached to the other end. Theoretically, if there is no attenuation by the fibre material, guided modes can propagate indefinitely. Note that not all the light launched into the fibre will experience total internal reflection.

7.6.2 Radiation Modes

Radiation modes occur when the light is launched at an angle of incidence less than the critical angle. Some of them propagate in the fibre core, and the remaining transmitted into the cladding. The radiation modes transmitted in the cladding propagate there by suffering reflection at the core-cladding boundary and even they may be transmitted back into the core so that mode coupling may occur with the higher order modes. Incident power will lost in the fibre due to radiation modes.

7.6.3 Leaky Modes

These modes attenuate by radiating power out of the core as it is propagating along the fibre. They are not totally internally reflected and their amplitude varies while propagating. The incident power is lost due to leaky modes also hence only a small portion will reach at the other end.

A mode remain guided as long as the propagation factor, β satisfies the relation

$$n_2 k < \beta < n_1 k$$

where, $k = 2\pi/\lambda$. The boundary between the guided mode and the leaky mode is defined by the cut-off condition $\beta = n_2 k$.

7.7
V-PARAMETER

The number of electromagnetic modes a fibre can support is described by dimensionless number called V-parameter. It is also called normalized frequency and is given by

$$V = \frac{2\pi a}{\lambda}(n_1^2 - n_2^2)^{1/2}$$
$$= \frac{2\pi a}{\lambda} n_1 \sqrt{2\Delta}$$

where a is the radius of the core, λ is the wavelength of light, n_1 and n_2 are the refractive indices of the core and cladding respectively. In general, a large number modes can propagate in a multimode fibre for a given value of V. When $V > 10$, the number of modes is approximately given by

$$N = \frac{V^2}{z}$$

For $V \leq 2.405$, the mode are said to be at cut-off condition. At cut-off, the optical power will in the cladding region is given by

$$P_{\text{clad}} = \frac{\Delta}{3\sqrt{N}} P$$

where P is the power in the optical fibre. The loss of power to the cladding increases as V increases.

The transferring of energy between the modes is called the mode mixing. Mode mixing lead to the loss of power, by coupling of lower-order modes with the higher order modes and then from higher order modes into the cladding modes. Thus, for high bandwidth transmissions, the number of modes in the fibre has to be restricted.

7.8
MULTIMODE AND SINGLE MODE FIBRES

In an optical fibre more than one electromagnetic mode can exist for any value of V-parameter greater than 2.405. A fibre that transmits many modes is called a multimode fibre. The total number of modes N that is supported for a step index fibre is given by

$$N \approx \frac{V^2}{z}$$

and that for a graded index fibre is

$$N = \frac{q}{q+2} = \frac{V^2}{z}$$

when $V \leq 2.405$ only one mode called the fundamental mode can propagate through the fibre (q, the profile parameter). An optical fibre that supported only the fundamental mode is called a single mode fibre. The wavelength at which the fibre becomes single mode is termed as the cut-off wavelength (λ_c). At cut-off $V_{\text{cut off}} = 2.405$ so that

$$\lambda_c = \frac{2\pi}{2.405} a(n_1^2 - n_2^2)^{1/2}$$

when the V-parameter increases above 2.405 the number of modes rise.

A single mode fibre has a small core diameter 5 to 8 μm and that for a multimode fibre is 50 to 80 μm. The cladding diameter is about 120–125 μm for both. Single mode fibres have many advantages as compared with the multimode fibres.

1. The losses due to attenuation, absorption scattering, model dispersion, etc. are low.
2. Attenuation is very small 0.15 dB/km.
3. Transmission bandwidth is very high ~1000 T bytes/km.
4. Good mechanical strength and temperature tolerance.

However, due to smaller core diameter mode coupling is more difficult in single mode fibres.

7.9
COHERENT OPTICAL FIBRE BUNDLE

In a fibre bundle, a large number of optical fibres are packed together. It can transmit an image from the input end to the output end in the form of bright or dark spots through each fibers in the bundle. In a coherent bundle, fibres are aligned with their relative position of the fibres at the input and output ends are the same (Figure 7.8). When such a bundle is illuminated at the input, the image in the same form will appear at the output. An important application of coherent optical fibre bundle is in medical/industrial endoscopes. By using it an illuminated interior portion can be viewed in the same form from outside. The resolutions is also very good since more than 10^5 are accommodated in a bundle of diameter of about 1 mm.

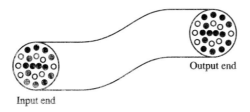

Figure 7.8 Coherent optical fibre.

If the fibres are not aligned, the relative position at the input and output are different, and is known as the incoherent optical fibre bundle. Here the output image will be scrambled (disordered). The image can be decoded by using a similar fibre in the reverse direction at the output end. An incoherent bundle can be used as a coder and the decoding is possible with the original bundle. Incoherent fibres have many applications such as illumination medical/industrial, electronic switching, calorimetry, display systems, photometric system, punch card reading, beam splitting/combining etc.

7.10
OPTICAL FIBRE CABLES

For practical applications optical fibres are incorporated in the form of cables. Fibre-optic cable has a protective enclosure surrounding a bare fibre. Its functions is to protect the fibre from any possible damage. The process of putting a fibre into such an enclosure is called cabling. These cables are grouped into loose type stranding, loose type high density, ribbon type, spoited core type, submarine type, etc. They may consist of single or many fibres and is available in rolls of 5 km length. For longer length requirements, they are joined together by using connectors. Optical fibre cables are used in different areas such as underground, submarine and sea bed, near by railway tracks, along with high tension electric lines, high temperature, chemical and radiation zones industries, buildings, etc. Different type of cables are used for each applications and the design consideration depends on the nature of application.

Optic fibre cables are designed such that they have certain common characteristics as follows:

(i) Minimum optical losses due to stress
(ii) High tensile strength
(iii) Protection against water vapour penetration, chemical reaction, animals etc.
(iv) Stability over the specified temperature range
(v) Low cable weight and volume
(vi) Ease of installation and handling
(vii) Low cost in installation and maintenance

Though design consideration is application depended, generally a cable consists of following elements as common:

Optical Fibre

It may be single or multiple number of fibres containing core cladding and acrylic coating. Multimode step index fibre, multimode graded index fibre and single mode step index fibre are available.

Buffer Material

Soft materials are used around the fibre to isolate it from radial compressions and localized stresses. There are two major type of buffer—In loose buffer the fibre is able to move within the buffering material, while responding to the stress. While in tight buffer the fibre is rigidly constrained to its position.

Strength Member

High tensile strength materials such as steel rod, copper wires Kelvar yarn, etc., are used to provide the longitudinal strength to the fibre.

Filler or Gel

It is added to take space between strength members and also to provide better buffering.

Outer Jacket

It provides protection to the fibre cable against chemical water, temperature animals, etc. It is made of polyethylene or silicon rubber.

A typical example of the aerial optical fibre cable structure is shown in Figure 7.9.

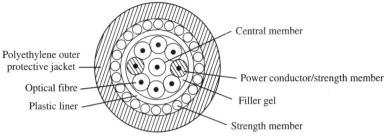

Figure 7.9 Cross-section of a typical aerial optical fibre cable structure.

Optical fibre cables are specified in terms of following parameters:
1. Transmission characteristics of the fibre wavelength range, mode structure, attenuation bandwidth, etc.
2. Maximum tensile loading specification and the operating temperature.
3. Number of fibres per tube and overall number of fibres.
4. Bending radius
5. Tension characteristics
6. Length and weight of the cable.
7. Uses of the fibre cable, etc.

7.11
OPTICAL FIBRE MATERIALS

In order to achieve better transmission characteristics selection of suitable materials for the manufacture of optical fibre is very important. Following requirements are considered in the material selection.
1. The material must be extremely pure and free from inhomogeneities so that the losses due to scattering, absorption and dispersion can be reduced.
2. In order to guide light the material must be transparent at the operating wavelength range.
3. The material must be capable to produce a refractive index profile in the step index or graded index fibres with appropriate doping.
4. The material must be capable to draw in the form of flexible, thin fibres of length about 5 km.
5. The material and fabrication cost must be low.

Glass and plastic are the most commonly used materials for core and cladding. In addition to it, shielding and cabling materials should have high mechanical strength and should be resistant to chemical reaction, temperature, moisture, etc.

7.12
GLASS FIBRES

Glass fibres are made from oxides of glass. The most important one is silica glass (SiO_2). It is a non-crystalline material with a well-defined melting point. It has a refractive index 1.458 at 850 nm. To obtain a small change in refractive index between the core and cladding suitable dopants are substituted to the silica. Dopants are added such that the refractive index of the core is slightly greater than that of the cladding. Dopants such as TiO_2, Al_2O_3, GeO_2, P_2O_5, etc., increase the refractive index of the silica. For example.

$$SiO_2 - GeO_2 \text{ core and } SiO_2 \text{ cladding}$$

Doping of silica with fluorine or B_2O_3 decreases the refractive index as in the fibre system.

$$SiO_2 \text{ core and } SiO_2 - B_2O_3 \text{ cladding}$$

The compound glasses used in the core and cladding must have similar thermal expansion coefficients, similar viscosities at the drawing temperature, low melting temperatures, long term chemical stability. The losses due to attenuation, dispersion and absorption etc. are high in multi-component glass fibres so that high quality fibres are made from high quality silica glass. Some important properties of glass fibres are as follows:

1. It can be used in the visible and infra red region.
2. It has low losses and high resistance to deformation, thermal expansion, chemical reaction, moisture etc.

7.13
PLASTIC FIBRES

Plastic fibres are either step index or graded index multimode fibres with polymer core (polymethylmethacrylate or perfluorinated polymer) and cladding. They are all plastic fibre known as Polymer Optical Fibre (POF). Since both core and cladding have similar properties their production process is simpler and involve low cost. They are more flexible than glass fibres and have a larger diameter (~1 mm) and numerical aperture. They are mechanically more easier to handle, plastic moulded connecters, splices etc., can be used efficiently.

The attenuation loss is high above 200 dB/km, and the bandwidth is as low as 5 MHz km. The transmission efficiency can reduce by repeated flexing and the operational temperature range is small (~50°C). Their use is limited to short distances and in the IR.

Plastic Cladded Silica (PCS) are also available, in which silica glass is used as core material and the cladding is made up of plastics. PCS fibres have higher losses and larger pulse dispersion than glass fibres, and are suitable to short distance applications.

7.14
RARE EARTHS DOPED OPTICAL FIBRES

These are known as active optical fibres in which rare earth elements are doped in the core. Commonly used materials are erbium and neodymium ions. They have a strong absorption and fluorescent line in the visible and infra red region. Such rare earth element doped optical fibres can perform amplification as the light is transmitting through it. To achieve efficient amplification the fibre is manufactured with reduced core diameter with rare earth ion concentration mostly along the central region. Erbium doped optical fibre is an optical fibre doped with erbium ions which act as an optical amplifier known as Erbium Doped Fibre Amplifiers (EDFA).

7.15
INFRA RED FIBRES

Infra red optical fibres are used to transmit optical radiations near IR to mid IR (10 μm) wavelength region. Materials such as GeO_2, GeS (chalcogenide glass), halides crystals (KCl), fluoride glass etc., are used to produce these fibres. Of these, fluoride glass fibres ZBLAN core/cladding and ZBLAN core/ZHBLAN cladding fibres have extremely low losses ~0.01 dB/km. Compositions of these fluoride fibres are given in Table 7.1. Note that ZHBLAN is formed by replacing some Hf for Zr in ZBLAN to decrease the index of refraction. The main drawbacks associated are the difficulty in producing in a longer length and divitrification of the glass.

Table 7.1 Compostion of ZBLAN halide glass fibre

	Composition %					
	ZrF_4	HfF_4	BaF_2	LaF_3	AlF_3	NaF
ZBLAN core	54	Nil	30	4.5	2.5	9
ZBLAN cladding	53	Nil	25	4	4	14
ZBLAN core	53	–	20	4	3	20
ZHBLAN cladding	39	13.3	18	4	3	22

The words ZBLAN/ZABLAN is formed by using first letters of the materials of the composition.

7.16
ZERO DISPERSION FIBRE

Single mode optical fibres in which pulse dispersion is zero are known as zero dispersion fibres. At zero dispersion wavelength the waveguide dispersion and material dispersion cancel each other. Material dispersion is due to the dependence of the refractive index of the fibre material and the wavelength. It also depends on the spectral width of the light source used. The waveguide dispersion is due to the dependence of the propagation constant on the wavelength and fibre parameters. The intermodel dispersion in graded index fibres may be reduced to zero level by choosing an optimum refractive index profile in the core region. Such fibres are known as dispersion modified fibres.

7.17
POLARIZATION MAINTAINING FIBRES

Single mode fibres capable of maintaining a single state of polarization are known as polarization maintaining fibres or single polarization single mode fibres.

7.18
ATTENUATION IN OPTICAL FIBRES

Attenuation represents the loss in an optical fibre as light propagate through it. It is an important parameter that determines how much light power can be received at the output end. An ideal fibre is expected to have no power or information loss at all. It determines the maximum transmission distance between the transmitter and the receiver that light can propagate without amplifications or repeters. Attenuation is wavelength dependent so that it is usually specified at a particular wavelength. For silica fibre the attenuation is minimum at 1550 nm (so it is used at this wavelength as the optimum operation wavelength).

As light propagates along the fibre, the power decreases exponentially with distance. Let $P(0)$ be the power at $z = 0$, then the power $P(z)$ after travelling a distance z can be expressed as

$$P(z) = P(0) \exp(-\alpha z)$$

where α is known as the attenuation coefficient. The fibre loss is expressed in decibels per kilometre i.e.

$$\alpha \text{ dB/km} = -\frac{1}{z} \log_{10} \frac{P_{(out)}}{P_{(in)}}$$

Her z is taken in km. For an ideal fibre $P_{out} = P_{in}$
$\therefore \quad \alpha = 0$ dB/km, i.e. 0 dB attenuation.
For a 3 dB attenuation.

$$3 \, (\text{dB/km}) = -\frac{1}{1} \log_{10}\left(\frac{1}{2}\right).$$

i.e., 50% of the input power will be reached at the output end.

Attenuation in optical fibres are due to several mechanisms. Some of them are as follows:
- Absorption
- Scattering effects
- Dispersion mechanisms
- Interface inhomogeneities, coupling losses, bending losses, etc.

The overall loss is just the sum of all the losses in dB due to individual effects in an optical transmission system.

7.18.1 Absorption Loss

Loss of optical power while light wave propagation along the fibre due to resonant absorption is the cause of absorption loss, and can take place by intrinsic and extrinsic constituents of the fibre material. Intrinsic absorption is caused by the absorption of light by one or more components of the glass material from which the fibre is formed, and occur in the specific wavelength regions of UV and IR. In the UV region it is due to the electronic absorption, when a photon interacts with an electron in the valence band, so that it excite to a higher energy level. IR absorption is associated with the vibration of chemical bonds between the atoms in the fibre material. The interaction of electromagnetic field with the vibrating bond results into a transfer of energy from the optical field giving rise to the absorption.

The intrinsic absorption represents a fundamental limit to the minimum attainable loss. It can occur even if the material used for manufacturing the fibre is free from density variation, in homogeneities, impurities, etc. The wavelength range where the optical fibre is operated, these absorption losses are insignificant. For example, in silica fibre at 1.5 μm absorption loss is very small (< 0.2 dB/km).

The extrinsic absorption is caused by the presence of transition metal ions (Fe, Cu, V, Co, Ni, Mn, Cr, etc.), as impurities and OH⁻ (hydroxyle) ions. The presence of hydrogen during the production of fibre or by corrosion of steel cable, strength members or by certain bacteria can produce hydroxyle ions. The presence of atomic defects or imperfections can also produce absorption losses. It can be reduced by using high quality materials, with impurities in the acceptable level by utilizing high quality manufacturing techniques.

7.18.2 Scattering Losses

Losses in an optical fibre can occur by scattering due to following effects:
1. Rayleigh scattering
2. MIE scattering
3. Non-linear scattering

Rayleigh Scattering

During the fabrication processes, in the optical fibre regions where the molecular density is higher or lower than the average density of the fibre can arise. In addition to this, compositional fluctuations in the oxides such as SiO_2, GeO_2, P_2O_5, etc., may also occur. These two effects can lead to variation in the refractive index along the fibre and can cause Rayleigh's type scattering of light. Rayleigh's scattering is proportional to $1/\lambda^4$, so that, the loss due to these type of scattering decreases with the increase in wavelength.

MIE Scattering

When the fibre contains inhomogeneities that are comparable in size (>$\lambda/10$) with the guided wavelength, scattering with large angular dependence may occur. The inhomogeneities can be imperfections in the glass materials such as stains, bubbles, diameter fluctuations, impurities in the core cladding interface, etc. This scattering is mainly in the forward direction and is known as MIE scattering.

Non-linear Scattering

In linear scattering, the scattered light has the same frequency as the incident light. While in non-linear scattering, the scattered light frequency shift is associated. The major non-linear scattering mechanisms are Brillion scattering and Raman scattering. Scattering due to these non-linear effects are very significant at high power density (optical power per unit area) i.e. when laser source or a fibre with very small core diameter fibre is used. Depending on the nature of interaction forward, backward or sideways scattering may occur.

Brillion and Raman scattering: In Brillion scattering and Raman scattering, the incident light non-linearly interacts with the fibre material to produce a phonon (quanta of lattice vibration) and scattered light.

Scattered light frequency is shifted by an amount ($\pm E_{ph}/h$) equal to that of the phonon frequency. Energy is conserved such that

$$h\nu_{scattered} = h\nu_{incident} \pm h\nu_{phonon}$$

Brillion scattering occurs in the backward direction (so that this effect reduces the light reacting the receiver). Raman scattering however, occurs in the forward direction, and the power is thus not reduced at the receiver). The Brillion scattering produces a lower energy phonon, whereas the Raman scattering produces a higher energy phonon.

7.18.3 Dispersion Mechanisms

A light pulse when propagates through an optical fibre, its shape get distorted in the time domain due to the dispersion (see Figure 7.10).

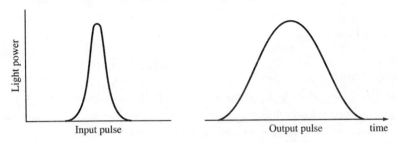

Figure 7.10 Pulse dispersion.

Non-ideal refractive index profile, non-uniform mode attenuation, mode mixing, bending, splicing, variation in optical power launching in various modes, etc., are some of the factors for these distortions. The amount of pulse spreading determine how close adjacent pulses are, and how close is a measure of information carrying capacity (or bit rate) of the optical fibre. The three prominent mechanisms of spreading out of pulse are:

1. Material dispersion
2. Waveguide dispersion
3. Model dispersion.

Material dispersion is caused by the variation of refractive index of the core with wavelength. The waveguide dispersion is due to the dependence of the propagation constant on wavelength. Both of these dispersions depend on the linewidth of the source. The finite spectral width of the light pulse effects such that the light with a larger wavelength travels faster as compared to the shorter wavelength, which result into the spreading out of the pulse. The intermodel dispersion is due to the delay as a result of variation in the group velocity of each mode.

7.18.4 Bending Losses

Radiative losses may occur, whenever a fibre bends, due to the condition for total internal reflection is not satisfied, resulting some light to escape from the core to the cladding. There are two types of bending losses due to macrobending and microbending (Figure 7.11).

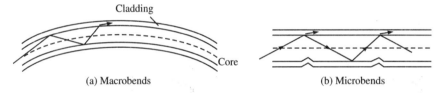

Fig. 7.11 Bending losses.

Macrobending Losses

Macrobending losses are due to large bends of the cable. As the radius of curvature decreases the losses increases exponentially. Manufactures usually specify certain bending radius at which the loss is minimum.

Microbending Losses

Microbends are the localized small scale variations in the radius of curvature in the fibre axis. They are usually formed during manufacturing or by non-uniform mechanical stresses while cabling.

7.18.5 Interface Inhomogeneities

Interface inhomogeneities such as impurities at the core, cladding or buffering interface or the geometric changes in the shape or size of the core during manufacturing can convert higher order modes into lossy modes in the fibre.

7.18.6 Joining Losses

Losses are introduced in a fibre by using connectors, couplers, splices, grating etc., in the network. These losses depend on many factors such as their design, properties of the fibre, misalignment, reflection, etc.

7.19
FIBRE CONNECTORS, SPLICES AND COUPLERS

An interconnect is an essential part in an optical fibre communication system, which is frequently needed at points such as to join an optical source to fibre, to provide intermediate fibre joints or links and to join fibre to receiver. Such joints are achieved by using interconnecting techniques. The use of a particular technique depends on whether a permanent or a removable connection is required. Every interconnect is subjected to a power loss, which depends on many factors, such as characteristics of the fibre – NA, fibre core/cladding size, refractive index profile, characteristics of the source and receiver, fibre length, fibre end qualities etc., and are developed with a minimum power loss.

A connector allows a demountable connection between fibre/or links or between source and fibre or fibre to a detector. A splice is a permanent or semipermanent joint between two fibres. They are used in long optical fibre communication network, where frequent connection/disconnection is not required. A coupler splits the optical power into many fibres or combine the optical power from several fibres into a single fibre.

7.20
FABRICATION OF OPTICAL FIBRES

A number of techniques have been used for the production of optical fibres. Two basic methods are direct-melt and vapour-phase oxidation processes. The direct-melt double crucible method is briefly described below.

7.20.1 Direct-melt Double-crucible Method

The double crucible facility consists of two platinum crucibles which have nozzles at the bottom (Figure 7.12). The inner crucible is used for taking the molten core, whereas the outer one is for the molten cladding glass. Appropriate temperature is maintained at the crucible by dc current or by heating in an rf (radio frequency) furnace. At first, core and cladding glasses of appropriate compositions are made in the form of glass rods by chemical processes or by melting mixtures from purified raw materials. These rods are then separately fed into the crucible to melt core/cladding glass from which the fibre is drawn through the nozzle. The orifice at the bottom is arranged such that the appropriate index grading is achieved by ionic diffusion between the core and the cladding glasses. The diameter of the core and cladding are maintained by the fibre pulling speed and the nozzle head of molten glass in each crucible. To obtain good quality fibre it is necessary to ensure that there is no contamination in the crucibles and also the glass materials are prepared in a clean environment using purified raw materials. Silica, chalgenide and halide glass fibres can be made using this direct-melt double-crucible technique. The advantages of this method are as follows:

1. The production rate is high (several hundreds of metres per minute)
2. The range of glass composition is larger. No restriction on diameter. Large scale production with low cost.

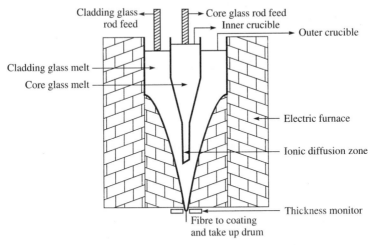

Figure 7.12 Direct-melt double-crucible method for fibre production.

Some of the other important fibre production techniques used are:
1. Outside vapour-phase oxidation
2. Vapour-phase axial deposition
3. Modified chemical vapour deposition
4. Plasma activated chemical vapour deposition
5. Internal chemical vapour deposition

7.21
FIBRE DRAWING PROCESS

The production of optical fibre consists of two major steps. The first step is to produce a rod of silica composition called preform. The preform consists of a core surrounded by a cladding with the desired refractive index profile. The second step is drawing into an optical fibre of desired size. For continuous mass production both are integrated.

In the fibre drawing process, the preform is attached to a precision feed which move into the drawing furnace. At this points, its tip is softened and starts to form optical fibre. A thickness monitor with feedback loop controls the diameter and the rate of drawing of fibre. A primary coating is applied immediately after it has been drawn to protect if from moisture, abrasion, etc., and a secondary coating is applied to resist mechanical strains. These coatings are cured by UV lamps or some other heating sources. Using pulling tractor assembly and take up drum at the bottom which is maintained at appropriate turning speed wound the fibres into reats.

116 Photonics: An Introduction

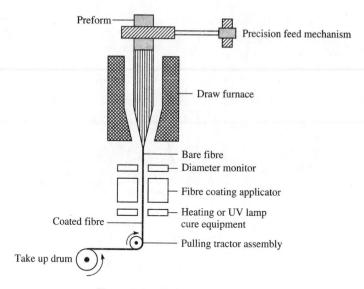

Figure 7.13 Fibre drawing process.

7.22

OPTICAL FIBRE COMMUNICATION SYSTEM

An optical fibre communication system (Figure 7.14), in its simplest form can be described as follows:

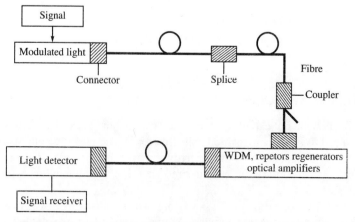

Figure 7.14 Schematic of the optical fibre communication system.

Light Source

Light from an optical source is modulated by the information signal. The modulation can be either external to the source or the source directly by varying the bias current. Semiconductor LEDs or laser diodes are used as the light source. Both analogue and digital form of signal can be used.

Optical Fibre

The modulated light is launched into the fibre, which guides from the transmitter to the receiver. The fibre has the property of low attenuation provided with good mechanical and environmental protection. Additional components such as connectors, couplers, splices, etc. may also be used while guiding the light.

Optical Signal Amplifers

If the link is long, optical signal may be attenuated or distorted because of absorption scattering, dispersion, etc. mechanisms of the glass materials. Optical amplifiers and regenerators are needed to reshape and amplify the signal for long distance communications.

Wavelength Division Multiplication

The use of wavelength division multiplexing technology which combines a number of wavelength in the same fibre and the use of erbium doped optical fibre amplifiers have increased the information capacity of the communication system.

Signal Receiver

Finally at the receiver point using photodiode, the information is detected and converted into electrical signal in the original format.

Optic fibre communication technology involves integration of waveguiding technology (optical fibre and integrated optics), semiconductor technology, electronic and communications technology and the developments in this area is very fast.

7.23
WAVELENGTH DIVISION MULTIPLEXER

The information carrying capacity of an optical fibre communication system can be enhanced considerably by transmitting many wavelengths simultaneously. Wavelength division multiplexer combines the light from many sources for transmission through the same fibre without affecting the individual signals. The optical demultiplexer separates the light into its component wavelengths. In brief wavelength division multiplexers and demultiplexers are the technologies that combine and separate signals at different wavelengths (Figure 7.15).

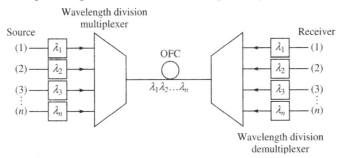

Figure 7.15 Functioning of wavelength division multiplexing/demultiplexing in an optical fibre communication system.

7.24
ADVANTAGES OF OPTICAL FIBRE COMMUNICATION SYSTEM

Important advantages offered by optical fibre communication system are as follows:
1. Wide bandwidth—several Tera Hz
2. Immunity to electromagnetic interference—optical fibres are non-conducting so that they will neither generate nor receive electromagnetic interference
3. No electromagnetic cross talk between channels
4. Light weight and low size
5. Low cost
6. No sparking
7. Compatibility with solid state technology
8. It provides a non-radiative means of information transferring to transfer large amount of data at a greater speed.

7.25
FIBRE OPTIC SENSORS

Many physical and chemical properties such as temperature, pressure, current, magnetic field, displacement, acceleration, flow rate, chemical concentration, pH value etc., can be measured by using optical fibres. Devices used for these applications are called fibre optic sensors and find applications in industry, automobiles, defense sector, etc. Basic principle of fibre sensor is shown in Figure 7.16.

Figure 7.16 Basic principle of fibre optic sensor.

Light from a LED or laser diode is passed to the zone where the parameter is to be sensed/measured. At this zone, some property of the light, intensity, frequency, phase, polarization or spectral distribution, etc., get modulated. The modulated light is then directed by the same or another fibre to the detector, and then processed. Sensors based on single mode fibres are much more sensitive than multimode fibres. Some of the fibre optic sensors are:

1. Temperature sensor
2. Pressure sensor
3. Displacement sensor
4. Rotation sensor (gyroscope)
5. Flow sensor
6. Fluid level sensor
7. Current sensor
8. Magnetostriction sensor
9. Mach–Zehndor interferometric sensor

Let us take an example of a temperature sensor. Piece of a semiconductor crystal GaAs is placed in between two fibre as shown in Figure 7.17. Light with a wavelength approximately equal to the band gap ($\lambda = hc/E_g$) is passed through the first fibre. Since the band gap of the GaAs is temperature dependent any change in temperature of the crystal will affect the absorption, so that the emergent light through another fibre can be utilized to measure the temperature.

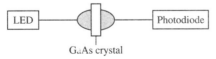

Figure 7.17 Fibre optic temperature sensor.

EXAMPLE 7.1 Refractive index of core and cladding of a doped silica fibre are 1.47 and 1.45, respectively. Calculate the following:
 (i) Critical angle for the core-cladding interfaces
 (ii) Maximum acceptance angle
 (iii) Numerical aperture
 (iv) Fractional refractive index.

Solution $n_1 = 1.47$, $n_2 = 1.45$

and
$$\theta_c = \sin^{-1}\left(\frac{n_2}{n_1}\right) = \sin^{-1}\left(\frac{1.45}{1.47}\right) = 80.5°$$

From Snell's law
$$n \sin \alpha_m = n_1 \sin (90 - \theta_c)$$
$$\sin \alpha_m = n_1 \sin (9.5)$$
$$\alpha_m \approx 14°$$
$$NA = \sin \alpha_m = 0.2425$$

and
$$\Delta = \frac{n_1 - n_2}{n_1} = 0.0136$$

EXAMPLE 7.2 Calculate the critical angle for the core-cladding interface and NA for a silica fibre with refractive indices 1.5 and 1.47 for core and cladding, respectively.

Solution
$$NA = (n_1^2 - n_2^2)^{1/2}$$
$$= 0.298$$
$$\theta_c = \sin^{-1}\left(\frac{n_2}{n_1}\right)$$
$$= \sin^{-1}\left(\frac{1.47}{1.50}\right)$$
$$= 78.5°$$

EXAMPLE 7.3 The relative refractive index change of a silicon fibre is 0.007, and the core refractive index is 1.46. Calculate the numerical aperture.

Solution
$$NA = n_1 \sqrt{2\Delta} = 1.46\sqrt{2 \times 0.007}$$
$$= 0.17$$

EXAMPLE 7.4 A multimode fibre operating at 800 nm has a core radius 30 μm, refractive index 1.46 and fractional refractive index of 0.009. Calculate the number of modes that can support by the fibre

Solution
$$V = \frac{2\pi}{\lambda} an_1 \sqrt{2\Delta}$$
$$= 46.1$$

where, λ = 800 nm, a = 30 μm, n_1 = 1.46 and Δ = 0.009

$$\therefore \quad N = \frac{V^2}{2} \approx 1063 \text{ modes}$$

EXAMPLE 7.5 A single mode fibre of radius 2.0 μm has a core refractive index 1.46 and fractional refractive index 0.015. Calculate the cut-off wavelength.

Solution
$$V = \frac{2\pi}{\lambda} an_1 \sqrt{2\Delta}$$

where, V = 2.405, a = 2.0 × 10^{-6} m, n_1 = 1.46 and Δ = 0.015

$$\therefore \quad \lambda = 1321 \text{ nm}$$

EXAMPLE 7.6 An optical power of 200 mW is launched into a fibre of length 0.25 km. At the receiving point the output power was measured to 10 μW. Calculate power loss in dB per kilometre.

Solution
$$\text{Power loss (dB/km)} = -\frac{1}{z} \log_{10} \frac{P_{out}}{P_{in}}$$
$$= -\frac{1}{0.25} \log_{10} \frac{200}{10}$$
$$= 5.2 \text{ dB/km}$$

REVIEW QUESTIONS

1. Explain the structure of an optical fibre.
2. Classify different types of optical fibres on the basis of their use.
3. Explain total internal reflection.
4. Draw the refractive index profile of optical fibres.
5. Explain refractive index profile of a step index and graded index fibre.
6. Define fractional change in index of refraction.
7. What are meridional and skew rays in optical fibres.
8. Define numerical aperture.

9. Discuss about the different types of modes in an optical fibre.
10. Define V-parameter. How is it related to the number of modes?
11. What are single mode and multimode fibres?
12. What are the advantages of single mode fibre over multimode fibres?
13. Describe a coherent bundle.
14. Explain the structure of an optical fibre cable.
15. What are the requirements to be considered in section of material for the manufacture of optical fibres?
16. Write short notes on the materials and for glass fibres, plastic fibres, rare earth doped fibres, infrared fibres.
17. What are zero dispersion fibres and polarization maintaining fibres?
18. Explain fibres loss in dB/km.
19. What are the different mechanisms of losses in an optical fibre?
20. Discuss different types of losses in an optical fibre.
21. Write short notes on fibre couplers, connectors and splices.
22. Describe the fabrication of optical fibre.
23. Describe an optical fibre communication system. What are its advantages?
24. Explain WDM.
25. What are optical fibre sensors? Give some examples.

8

Modulation of Light

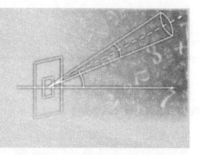

8.1
INTRODUCTION

Modulation is a process by which the waveform of a high frequency carrier wave is suitably modified to transmit information. One of the parameters associated with the carrier wave, such as amplitude, irradiance, frequency phase or polarization—is varied to enable it to carry information. Modulation process has been classified as analog or digital. In analog modulation, the information signal varies with the high frequency carrier wave in a continuous manner. In digital modulation, discrete changes in the intensity of the carrier wave are caused by the information signal. Information is then transmitted by the high frequency signal as a series of discrete pulses (0 and 1).

8.2
DIRECT AND EXTERNAL MODULATIONS

8.2.1 Direct Modulation

To modulate optical signals in devices using light emitting diodes and semiconductor diode lasers, two schemes are used.

In direct modulation the current injected to the light source is modified using an electronic circuit. The output light is controlled by the injected current to achieve desired modulation. It has the following advantages and disadvantages.

Advantages

1. Circuit is relatively simple and compact.
2. Device speed is controlled by internal process such as e-h recombination rate, cavity length, etc.

Disadvantages

1. Emission frequency can alter as the drive current is changed.
2. Only the output intensity can be modulated easily. For phase and frequency modulation additional care should be needed in electronic circuit design.

8.2.2 External Modulation

In external modulation scheme, the light output passes through a material whose optical properties can be modified by an external means. Depending on the mechanism used, one can have electro optic, acousto optic or magneto optic modulators.

Advantages

1. Device speed can be controlled by modulator properties and are quite fast.
2. Emission frequency is unaltered.
3. Phase can be altered.
4. Modulator part is quite big and difficult to fine tune.

8.3
DIRECT MODULATION OF LED

A simple circuit for direct modulation of LED is shown in Figure 8.1; in which modulation is achieved by varying the drive current. It is also called current modulation. The time-varying input signal is applied about the bias point kept at the linear portion of the output characteristics.

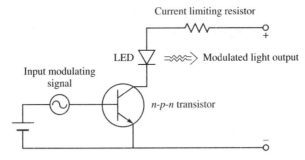

Figure 8.1 Direct modulation in LED.

To provide sufficient gain to the signal an *n-p-n* transistor is connected is as shown in Figure 8.1 and the signal is applied to the base. The current through the LED is limited by the limiting resistor. The variation of the base current (signal) varies the current flow and hence the forward injection current of LED so that a modulated output is obtained (Figure 8.2). The amplitude of the signal must be kept small enough to avoid the saturation portion of the transfer characteristics to avoid

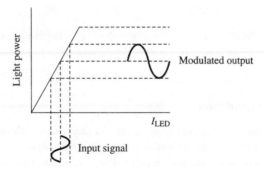

Figure 8.2 Direct modulation in LED.

distortion of signal. When the current is operated in the linear range of LED and the transistor, the light output of the LED is directly proportional to the input signal applied in the base circuit.

The frequency response is determined by the following factors

1. Doping level in the active region
2. Injection carrier lifetime in the recombination region
3. The parasitic capacitance.

Under constant forward bias, the delay produced by parasitic capacitance is negligible.

8.4
DIRECT MODULATION IN SDL

Semiconductor diode laser is a very important source for optical communication system, in which for the transmission of information the optical output of the laser diode has to be modulated. It can be modulated directly by varying the injection current. When the drive current is modulated at high frequency, it requires that the laser should react to it, so that the light output follows the current drive pulse accordingly. The most important and the simplest approach to modulation of SDL is direct modulation in which the current through the laser is modulated (Figure 8.3).

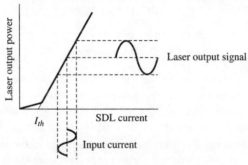

Figure 8.3 Direct modulation in SDL.

The possibility of monolithic integration of SDL, and the modulating electronic circuit, the problem of high speed modulation of laser output is one of great technologically important field of research. Depending on the application, the modulation schemes used are large signal modulation, small signal modulation and pulse code modulation.

The overall modulation response of SDL is controlled by both extrinsic and intrinsic factors.

8.4.1 Extrinsic Factors

Laser Heating

A high current is necessary to operate the device at high speed. When the laser is biased at a high current, laser heating will be produced. The heat produced can produce non-linearities in the device parameters such as gain profile. Two important non-linearity effect are harmonic and intermodulation distortions.

Biased at High Power

High power produces a failure of the laser due to the damage produced in the mirror of the cavity. Thus, the laser has an upper limit on injection at which it can be operated safely.

Extrinsic Parasitics

The laser drive circuit must be designed such that the resistance, capacitance, inductance, etc., do not limit the device response.

8.4.2 Intrinsic Factors

Intrinsic modulation limitation factors include all the factors that limits the gain and speed of modulation, which arise from cavity design, carrier drift, diffusion, etc.

8.5 EXTERNAL MODULATORS

Block diagram of external modulation is shown in Figure 8.4. It consists of a light source (SDL or LED). The light is then modulated by means of modulating device by utilizing electro optic, acousto optic or magneto optic effects.

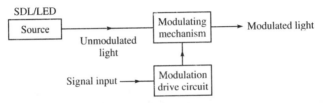

Figure 8.4 External modulation of light.

8.5.1 Electro Optic Effect

In certain crystals in the presence of an electric field, there is a change in the index of refraction (birefringence) is observed. This phenomenon is known as electro optic effect. When an electric field is applied to such crystals under certain geometrical orientations, the applied field acts differently on two linearly polarized light waves passing through the crystal, and this can introduce a change in the refractive index. A change in the direction of the ordinary and extra-ordinary ray is caused that is proportional to the electric field. The effect of applied field depend on the crystal structure and symmetry of the material.

Electro optic effect is commonly used in light modulators, where the change in refractive index (RI) is utilized for amplitude, phase or frequency modulation. The application includes

1. Q-switching and mode locking in lasers.
2. Couplers, switches, bistable devices, etc. in integrated optics.
3. Generation of high peak power optical pulses.
4. Impression of information on to optical beams.

If the field dependent change in the refractive index is linearly proportional to the electric field, corresponding electro optic effect is called as Pockel effect or linear electro optic effect. If the change in the indices of refraction is proportional to the square of the applied electric field the effect is called as Kerr effect or quadratic electro optic effect. Some examples for electro optic materials are as follows:

(i) Potassium dihydrogen phosphate KDP (KH_2PO_4)
(ii) Potassium didutarium phosphate KD^*P (KD_2PO_4)
(iii) Ammonium dihydrogen phosphate (ADP) ($NH_4H_2PO_4$)
(iv) Quartz (SiO_4)
(v) Lithium niobate ($LiNbO_3$)
(vi) Gallium arsenide (GaAs)
(vii) Cadmium telluride (GdTe)
(viii) Gallium phosphide (GaP)
(ix) Lithium tantalate ($LiTaO_3$)
(x) Zinc sulphide (ZnS)

8.5.2 Pockel Effect in KDP Crystal

KDP is a uniaxial (one optic axis) crystal in the absence of electric field (Figure 8.5). When a light wave is passed through the optic axis which is assumed to be lies along the z-axis, it travels along the same direction (z-axis) with two polarizations in the x and y direction as a principal wave with refractive index n_0. When an external electric field (E_z) is applied along the z-direction the crystal becomes biaxial (two optic axes), the principal axes x and y rotated through

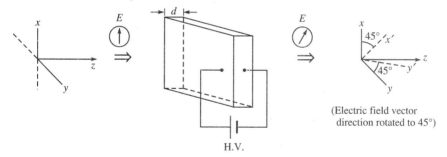

Figure 8.5 Pockel effect in KDP crystal.

45° into new principal axes x' and y'. For a wave propagation along the z-direction, the refractive indices for waves polarized along x' and y' are given by

$$n_{x'} = n_0 - \frac{1}{2} n_0^3 \gamma E_z$$

and

$$n_{y'} = n_0 + \frac{1}{2} n_0^3 \gamma E_z$$

Here γ is a coefficient describing electro optic effect in this configuration.

The magnitude of the change in refractive index produced in the presence of electric field is given by

$$\Delta n = \frac{1}{2} n_0^3 \gamma E_z$$

If V is the voltage applied across the crystal of thickness d, then

$$E_z = \frac{V}{d}$$

The change in phase introduced between two polarized beams when the light propagate a distance l through the crystal is given by

$$\Delta \phi = \frac{2\pi}{\lambda} \Delta n l = \frac{\pi V n_0^3 \gamma l}{\lambda d}$$

Note that though the refractive index change is low, the phase shift produced for a path ~1 cm and a field n 10^4 V/cm, is considerably large.

8.6

PRINCIPLE OF PHASE MODULATION (ELECTRO OPTIC RETARDATION)

Consider an optical field $[E = E_0 \exp(i\omega t)]$ propagating along z-direction incident at $z = 0$, normal to the plane of a KDP crystal (Figure 8.5). The optical field can be resolved into two mutually orthogonal components.

$$E_x = E_y = \frac{E_0}{\sqrt{2}} e^{i\omega t}$$

which represents a wave with the polarized components along the two principal axes x and y directions. When an electric field is applied along the z-axis of the crystal, a small change in the refractive index is observed by Pockel effect, the two waves travel with different propagation constants according to

$$E_{x'} = \frac{E_0}{\sqrt{2}} \exp i\left(\omega t - \frac{2\pi}{\lambda} n_{x'} z\right)$$

and

$$E_{y'} = \frac{E_0}{\sqrt{2}} \exp i\left(\omega t - \frac{2\pi}{\lambda} n_{y'} z\right)$$

If l is the length of the medium, the phase difference between the two components at the output plane is given by

$$\phi_{y'} - \phi_{x'} = \frac{2\pi}{\lambda}(n_{y'} - n_{x'})l$$

$$\Delta\phi = \frac{2\pi}{\lambda}\left(n_0 + \frac{1}{2}n_0^3 \gamma E_z - n_0 + \frac{1}{2}n_0^3 \gamma E_z\right)l$$

$$= \frac{2\pi}{\lambda} n_0^3 \gamma E_z l = \frac{2\pi}{\lambda} n_0^3 \gamma V$$

The net phase difference ($\Delta\phi$) between two components resulting from the application of voltage is known as electro optic retardation. This is the principle used for phase modulation using linear electro optic effect.

If the applied voltage is oscillatory with a frequency ω_m, represented by

$$V = V_0 \sin \omega_m t$$

which leads to a sinusoidal variation of the output light. The output beam contains in addition to the wave at the fundamental frequency ω, various side bands at frequencies $\omega \pm \omega_m$, $\omega \pm 2\omega_m$ ... etc., which can be seen using Fabri-Perot interferometer.

8.7

ELECTRO OPTIC AMPLITUDE MODULATOR

Figure 8.6 shows the arrangement for the small signal amplitude modulation with electro optic effect. We know that the superposition of two linearly polarized waves leads to a circularly polarized light (if the phase difference between them is $\pi/2$ and of equal amplitude). The incident light is passed through a vertical polarizer. A quarter wave plate introduced between the vertical polarizer and the modulator introduces a phase difference of $\pi/2$ between the two polarized components. The phase retardation between the components polarized along the new optic axes (x' and y') can be introduced by an external electric field, and the magnitude of retardation is proportional to the electric field. The elliptically polarized beam from the crystal is then passed through an analyzer oriented horizontally to the input polarization, then the amplitude of the beam from the analyzer will be modulated.

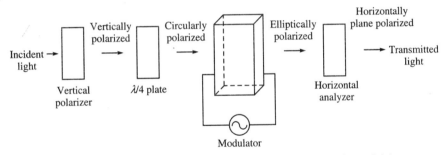

Figure 8.6 Arrangement for small signal electro optic amplitude modulator.

For small voltages V, the transmittance is nearly linear with the applied voltage and is given by

$$\frac{I}{I_0} = \frac{1}{2}\left(1 + \sin \pi \frac{V}{V_\pi}\right)$$

where

$$V_\pi = \frac{\lambda}{2\gamma n_0^3}$$

is the voltage required for maximum transmission or to produce a phase difference of π. It is also called as half-wave voltage. V_π depends on the electro optic material and the wavelength.

For small signal modulation, the modulator is usually biased at some point say Q, and a small signal modulating voltage is superimposed. At the biasing point, the transmission characteristics is almost linear (Figure 8.7).

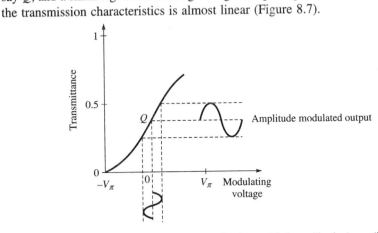

Figure 8.7 Principle of small signal amplitude modulation with electro optic effect.

As the electric field is applied along the direction of propagation of the light they are called longitudinal electro optic modulators. They have two disadvantages as follows:

(i) The contacts are made at the ends in the longitudinal direction which limits the amount of light.

(ii) Semitransparent contacts will lead to non-uniform transmission and losses.

In transverse electro optic modulators, the electric field is applied normal to the direction of propagation.

The finite transit time (time taken by the light to traverse the crystal) limits the maximum modulation frequency, which is given by

$$(v_m)_{max} = \frac{c}{4nl}$$

For a KDP crystal with $l = 1$ cm and $n = 1.5$, we get

$$(v_m)_{max} = 1.5 \text{ GHz}$$

Limitation due to such transition time can be overcome by applying electric field in such a way that it propagates along the crystal. Such modulators are called travelling wave modulators.

8.8
KERR MODULATORS

In Kerr modulators, the change in refractive index between light polarized parallel and perpendicular to the induced optic axis is proportional to the applied electric field, i.e.,

$$\Delta n = K\lambda E^2$$

where λ is the wavelength, K is the Kerr constant. Many media (solids and liquids) such as nitrobenzene, potassium tantalate niobate, barium titanate, ferroelectric materials exhibit Kerr effect and are used for the Kerr modulator. In a simple set up, one of the liquid is taken in a Kerr cell placed between two crossed polaroid and applied with the electric field can be used as a modulator.

8.9
MAGNETO OPTIC MODULATION

Magneto optic modulators are based on magneto optic effect (also known as Faraday effect). It is the property of some transparent materials that rotate the polarization when light is propagated through such substances which is subjected to a magnetic field. The rotation of the plane of polarization is given by

$$\theta = VBL$$

where V is the constant of proportionality called the Verdet constant, B is the magnetic flux applied parallel to the direction of propagation and L is the length of the material. Basically, it concerns with the change in refractive index for the polarized wave along two directions when subjected to the magnetic field, so that the intensity modulation of light beam can be obtained.

8.10
ACOUSTO OPTIC EFFECT

Acousto optic effect is the induced change in the refractive index of an optical medium caused by mechanical strains produced by an acoustic wave.

A sound wave produces a sinusoidal variation of density—i.e. regions of pressure maxima and minima—of the material. The change in density causes a change in the index of refraction which varies periodically with a wavelength equal to that of the acoustic wave in the medium. Hence, the periodically refractive index varying medium acts as a refractive index grating.

When light incident on such a refractive index grating diffraction takes place. At low acoustic frequencies usually multiple order diffraction are observed, which is known as Raman–Nath diffraction, and at high acoustic frequencies only single order diffraction is produced which is known as Bragg diffraction. Acousto optical phenomena have been used in many applications such as acousto optic modulators, mode locking and Q-switching lasers, deflectors, frequency shifters, etc.

Acousto Optic Modulator

The principle of acousto optic modulation is schematically shown in Figure 8.8. A beam of monochromatic light is allowed to incident upon a medium, which is subjected to mechanical strain by means of an appropriate transducer. The acoustic waves produced will change the refractive index of the medium periodically. As the light enters the medium, the wavefront close to the pressure maxima (compression) will encounter a high refractive index so that it advances with a lower velocity than those portions of the wavefront which encounter pressure minima (lower refractive index region) so that advances with higher velocity. Consequently the plane (wavefront) acquires a wavy appearance as they propagate through the medium. Because of this refractive index grating, the beam passing through it will be diffracted. The intensity of the diffracted light at small acoustic frequencies is proportional to the intensity of the acoustic wave, which is the principle used in acoustical modulation of optical signal.

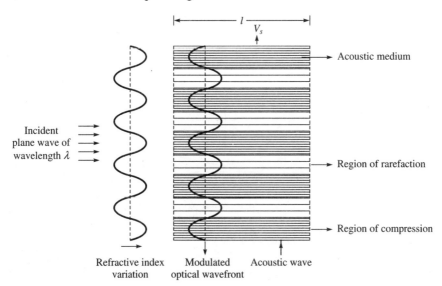

Figure 8.8 Acousto optic modulator using Raman–Nath diffraction and Bragg diffraction.

Many materials solids, liquid and gases exhibit acousto optic effect, some of them are, KDP, ADP, GaAs, CdS, Quartz, flint glass, water, kerosene, etc.

The fraction of the power of the incident optical beam transferred into the diffracted beam is given by

$$\frac{I_o}{I_{in}} = \sin^2 \frac{\pi l}{\sqrt{2\lambda}} \sqrt{\frac{p^2 n^6}{\rho V_s^2}} I_a$$

where

λ is wavelength of the light wave
n is refractive index of the medium
ρ is density of the medium
l is optical path in acoustic medium
p is photo elastic constant of the medium
I_a is acoustic wave intensity
V_s is velocity of the acoustic wave

The acousto optic modulator is operated with a few volts, and is relatively less expensive than electro optic modulators which require a relatively high voltage in the range of kilovolts. Acousto optic modulators have a bandwidth up to 50 MHz.

The medium is taken in a cell, in which acoustic waves are produced by a piezo electric transducer applying an oscillating voltage. The periodic compression and rarefaction will produce a periodic refractive index variation. The reflected acoustic waves at the other ends are eliminated by placing suitable absorbers. Monochromatic light wave is passed through the cell in a direction perpendicular to the acoustic wave. The output is observed on a screen on which a direct beam with the diffracted spots on both side can be viewed. With suitable acoustic power levels at low acoustic frequencies Bragg diffraction pattern can be observed which contains only one order (Figure 8.9). At higher acoustical frequencies Raman–Nath diffraction can be observed which consists of a multiple order diffraction pattern as shown in Figure 8.10.

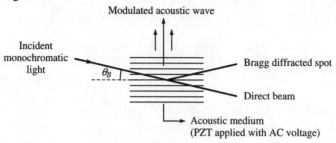

Figure 8.9 Bragg diffraction.

Note that in the case of Bragg diffraction, the incident light wave falls at the Bragg angle θ_B.

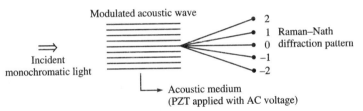

Figure 8.10 Raman–Nath diffraction.

EXAMPLE 8.1 Electro optic coefficient of a KDP crystal of wide 10 mm is 10.3×10^{-12} m/V. Calculate the change in refractive index when a voltage 4 kV is applied $n_0 = 1.5074$.

Solution
$$\Delta n = \frac{1}{2} \gamma n_0^3 E_z$$
$$= \frac{1}{2} \times 10.3 \times 10^{-12} \times (1.5074)^3 \times \frac{4000}{10 \times 10^{-3}}$$
$$= 7.056 \times 10^{-6}$$

EXAMPLE 8.2 In a lithium niobate crystal with the electro optic coefficient 30.8×10^{-12} m/V and a refractive index 2.29, a voltage of 10^4 V is applied across the crystal. Calculate the change in the refractive index within the crystal, if it has a dimension $1 \times 1 \times 1$ cm^3.

Solution
$$\Delta n = \frac{1}{2} n_0^3 \gamma \frac{V}{d}$$
$$= \frac{1}{2} \times 30.8 \times 10^{-12} \times (2.29)^3 \times \frac{10^4}{1 \times 10^{-2}}$$
$$= 1.849 \times 10^{-4}$$

EXAMPLE 8.3 What voltage has to be applied to an ADP crystal, with electro optic coefficient 8.56×10^{-12} m/V and refractive index 1.5266 to produce a phase shift of π between the two polarized beams? The free space wavelength is 0.546 µm.

Solution
$$\Delta \phi = \frac{2\pi}{\lambda} \gamma n_0^3 V$$
$$\pi = \frac{2\pi}{0.546 \times 10^{-6}} \times 8.56 \times 10^{-12} \times (1.5266)^3 V$$
$$\therefore \quad V = 8964 \text{ V}$$

EXAMPLE 8.4 Calculate the phase shift produced to one of the polarized component due to electro optic effect.

$\gamma = 10.5 \times 10^{-12}$ m/V;
$V = 10^6$ V/m;
$n_0 = 1.512$;
$l = 1 \times 10^{-2}$ m;
$\lambda = 500$ nm.

Solution
$$\Delta\phi = \frac{2\pi}{\lambda}\frac{1}{2}\gamma n_0^3 E_z l$$
$$= \frac{2\pi}{500 \times 10^{-9}} \times \frac{1}{2} \times 10.5 \times 10^{-12} \times (1.512)^3 \times 10^6 \times 1 \times 10^{-2}$$
$$= 0.72\pi \quad \text{radians}$$

EXAMPLE 8.5 Calculate the half-wave voltage of a KDP crystal at $\lambda = 632.8$ nm, with $n_0 = 1.512$ and $\gamma = 10.5 \times 10^{-12}$ m/V.

Solution
$$V_\pi = \frac{\lambda}{2n_0^3 \gamma}$$
$$= \frac{632.8 \times 10^{-9}}{2 \times (1.512)^3 \times 10.5 \times 10^{-12}}$$
$$= 8.717 \text{ kV}$$

EXAMPLE 8.6 Calculate the maximum electro optic modulation frequency for a KDP crystal of length 0.5 cm and $n_0 = 1.512$.

Solution
$$(v_m)_{max} = \frac{c}{4n_0 l}$$
$$= \frac{3 \times 10^8}{4 \times 1.512 \times 0.5 \times 10^{-2}}$$
$$= 9.921 \text{ GHz}$$

REVIEW QUESTIONS

1. What is mean by modulation of light?
2. Explain direct and external modulation of light. What are its merits and demerits?
3. Explains the direct modulation in LED.
4. Discuss the direct modulation in SDL.
5. What is electro optic effect?
6. Describe the Pockel effect in KDP crystal.
7. Describe the phase modulation in KDP crystal.
8. Describe the electro optic amplitude modulation in KDP crystal.
9. What is half wave voltage?
10. What is the principle of Kerr modulation?
11. What is the principle of magneto optic modulation?
12. What is acousto optic effect?
13. Explain the principle of acousto optic modulator.
14. Describe the Raman–Nath and Bragg diffraction.

9
Non-linear Optical Processes

9.1
INTRODUCTION

When a light wave propagates through an optical medium, the oscillating electromagnetic field can exert a polarizing force on the electrons comprising the medium, i.e. the oscillating electric field associated with the light beam can act as a small perturbation on the atoms. For weak optical intensities such as that in ordinary light, in a linear dielectric medium the polarization produced is proportional to the electric field associated with the light wave (as shown in Figure 9.1).

$$P = \varepsilon_0 \chi E$$

where P is the polarization, χ is the dielectric susceptibility, E is the electric field and ε_0 is the absolute permeability.

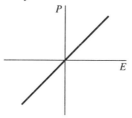

Figure 9.1 Relation between P and E in a linear dielectric medium.

If the electric field of the radiation is comparable with the atomic fields ($\sim 10^8$ V/cm) then the relation between the polarization and the electric field is non-linear as in the Figure 9.2.

In an isotropic medium, the general relation between P and E can be expressed as

$$P = \varepsilon_0(\chi^{(1)}E + \chi^{(2)}E^2 + \chi^{(3)}E^3 + \ldots)$$

In this expansion, $\chi^{(1)}$ is the linear susceptibility. It is much larger than the non-linear coefficient $\chi^{(2)}$, $\chi^{(3)}$, etc. (Note that in general $\chi^{(i)}$ is a tensor quantity in an anisotropic medium). Since lasers have enormous light power, the light fields

135

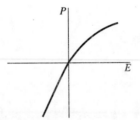

Figure 9.2 Relation between P and E in a non-linear optical medium.

needed to generate such non-linearity can be achieved by using high power laser sources. A medium in which the polarization is described by a above non-linear equation is called a non-linear medium.

9.2
SECOND HARMONIC GENERATION

Second Harmonic Generation (SHG) (also called frequency doubling) is a non-linear optical phenomena in which a part of energy of a high power optical beam of frequency ω when propagating through a certain non-linear optical medium (eg. KDP) is converted to a wave of frequency 2ω.

Suppose that the field incident on the non-linear medium has the form

$$E = E_0 \cos \omega t$$

Then the polarization is given by

$$P = \varepsilon_0 \chi^{(1)} E_0 \cos \omega t + \varepsilon_0 \chi^{(2)} E_0^2 \cos^2 \omega t + \varepsilon_0 \chi^{(3)} E_0^3 \cos^3 \omega t + \cdots$$

$$= \varepsilon_0 \chi^{(1)} E_0 \cos \omega t + \varepsilon_0 \chi^{(2)} E_0^2 \left(\frac{1 + \cos 2\omega t}{2} \right)$$

$$+ \varepsilon_0 \chi^{(3)} E_0^3 \left(\frac{\cos 3\omega t + 3 \cos \omega t}{4} \right) + \cdots$$

$$= \frac{1}{2} \varepsilon_0 \chi^{(2)} E_0^2 + \varepsilon_0 \left(\chi^{(1)} + \frac{3}{4} \chi^{(3)} E_0^2 \right) E_0 \cos \omega t + \frac{1}{2} \varepsilon_0 \chi^{(2)} E_0^2 \cos 2\omega t$$

$$+ \frac{1}{4} \varepsilon_0 \chi^{(3)} E_0^3 \cos 3\omega t + \cdots$$

The first term represents dc polarization across the medium, the second term which oscillates at ω represents the polarization due to fundamental, the third term which oscillates at a frequency 2ω is called the second harmonic polarization, the fourth term is called the third harmonic polarization and so on. This is shown schematically in Figure 9.3. In the case of non-centro-symmetric materials (anisotropic crystals) both quadratic and cubic terms are present. However, generally the third harmonic term is substantially smaller than the second order term.

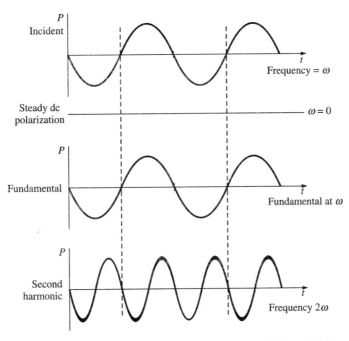

Figure 9.3 Non-linear polarization in the presence of high light field.

A schematic of the experimental set up to observe the second harmonic generation is shown in Figure 9.4. A beam from a laser with sufficient power (say Ruby laser at $\lambda = 694.3$ nm) is focussed into a quartz crystal. A part of the incident light energy is converted into second harmonic wave which has a wavelength 347.1 nm. The two beams—fundamental and second harmonic—are then separated by a prism and detected on a photographic plate.

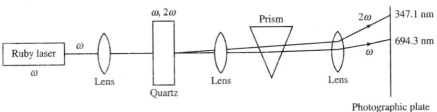

Figure 9.4 Experimental set up for SHG.

The following points has to be noted with the SHG:
1. Only those crystals that lack inversion symmetry exhibit SHG.
2. Condition for efficient SHG is that the propagation vector for the second harmonic radiation required to be equal to twice that of the incident beam i.e.
$$k^{2\omega} 2\omega = 2k^{\omega} \omega$$
This is known as phase matching criteria.

Using $k^\omega = \omega\sqrt{\mu\varepsilon n^\omega}$, it follows:
$$n^{2\omega} = n^\omega$$
The indices of refraction at the fundamental and second harmonic frequencies must be equal.

3. The irradiance of the second harmonic reaches a maximum after the wave propagated a distance
$$l_c = \frac{2\pi}{\Delta k} = \frac{2\pi}{k^{2\omega} - 2k^\omega}$$
$$= \frac{\lambda}{2(n^{2\omega} - n^\omega)}$$

l_c is known as coherence length, which is a measure of the maximum crystal length that is useful in producing the second harmonic power.

4. The second harmonic generation efficiency is proportional to the intensity of the fundamental wave.

Important application of SHG is that it enables to extend the range of laser wavelength in the blue and UV part of the spectrum.

9.3
SELF-FOCUSSING AND DEFOCUSSING

The dependence of refractive index of a medium and the intensity of the light beam can give rise to a non-linear phenomena namely self-focussing or defocussing of the light beam. When a light beam with Gaussian intensity profile incident on a medium (such as carbondisulphite, acetone, etc.) the dependence of refractive index on the electric field of the light wave can be expressed as
$$n = n_1 + n_2 E_0^2$$
where n_1 is the refractive index of the medium in the absence of light and n_2 is a constant representing self-focussing/defocussing effect. When the beam interacts, it creates a refractive index gradient in the medium. If this interaction is such that $n_2 > 0$, the refractive index is maximum at the axis and decreases away from the axis. Thus the medium acts as a converging lens so that the incident beam get self focussed [Figure 9.5(a)]. On the other hand, if $n_2 < 0$; the refractive index is minimum at the axis and increases as move away from the axis so that the beam will undergo defocussing effect [Figure 9.5(b)]. The spreading of light due to diffraction effect can also be approximated as the diverging lens. If non-linear effects dominate than the diffraction divergence effect the beam will be self-focussed.

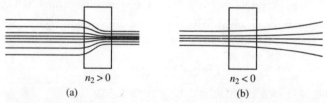

Figure 9.5 (a) Self-focussing effect, (b) Self-defocussing effect.

9.4
OPTICAL PARAMETRIC INTERACTIONS

Optical parametric amplification is a non-linear optical phenomena, in which a strong pump wave at ω_3 in a non-linear crystal causes a simultaneous generation of radiation at ω_1 and ω_2 where $\omega_3 = \omega_1 + \omega_2$. It involves the transfer of power from the pump wave at a higher frequency ω_3 to wave at lower frequencies ω_1 and ω_2 (Frequency down conversion). $\omega_1 = \omega_2$ is a special case which is the exact reverse of the second harmonic generation.

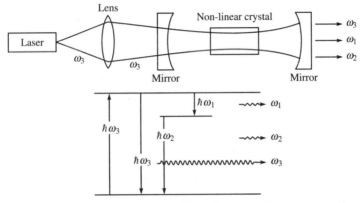

Figure 9.6 Optical parametric oscillator (frequency down conversion).

Parametric interaction in a non-linear crystal can be used to covert an optical signal at a low frequency ω_1 to a high frequency ω_3 by mixing it with a pump laser beam at ω_2 such that

$$\omega_3 = \omega_1 + \omega_2$$

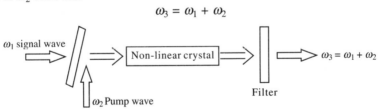

Figure 9.7 Frequency up conversion.

In the experimental set up, the optical signal at ω_1 and a strong pump beam at ω_2 are combined in a non-linear optical crystal to generate a beam at

$$\omega_3 = \omega_1 + \omega_2$$

9.5
FOUR WAVE MIXING

In four wave mixing, two counter propagating coherent pump beams are allowed to interact with a third input wave in a non-linear optical medium which results into the generation of a phase conjugated wave due to the third order non-linear polarization. In degenerate four wave mixing all the four waves are at same frequency with plane wavefront, with the pump beams represented by $k_1(\omega)$ and $-k_2(\omega)$, and the incident wave by $k_3(\omega)$, then for the phase conjugate wave

$$k_4(\omega) = k_1(\omega) - k_2(\omega) - k_3(\omega)$$
$$= -k_3(\omega)$$

For a phase conjugate waves, the phase at each point in space is exactly same to that of the incoming wave and exactly retraces the original path.

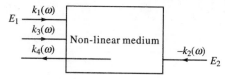

Figure 9.8 Four wave mixing.

9.6

MULTIPHOTON ABSORPTION

When a light beam interacts with an absorbing medium, the energy absorbed is given by

$$E = N\hbar\omega$$

This energy is equivalent to bandgap or work function of the material. Here N is the number of photons involved. Usually with ordinary light a single photon is absorbed. However with a high power laser beam more than one photon can be absorbed simultaneously which is known as multiphoton effect. A special case is a two photon absorption, shown in Figure 9.9. Here two photonsc of energy $\hbar\omega/2$ is absorbed to excite an atom from a state E_1 to an upper state of energy $E_2 = E_1 + \hbar\omega$. A notable thing is that the probability of absorption is proportional to the square of the intensity of the light beam

$$i \propto I^2$$

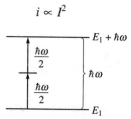

Figure 9.9 Two photon absorption

REVIEW QUESTIONS

1. Explain what is meant by non-linear optical phenomena.
2. Explain second harmonic generation.
3. Explain self-focussing phenomena.
4. Explain frequency up/down conversion by parametric oscillation.
5. What is four wave mixing?
6. Explain multiphoton absorption.

10
Integrated Optics

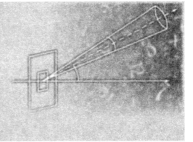

10.1
INTRODUCTION

Integrated optics is the monolithic integration of necessary optical and electrical components to function it as a device. By means of monolithic integration all the components are fabricated on a single semiconductor chip called the optical IC (Integrated Circuit) which contains miniaturized electrical and optical components on a thin film waveguide structure provided with necessary interconnects. These components have dimensions usually of the order of wavelength and within the network, light is guided between the components in the waveguide structure. By a proper choice of active and passive components and the design, it is possible to perform generation, modulation, amplification, coupling filtering, multiplexing, detection, etc. of the optical wave. The integration enables to do as much signal processing as possible directly on the optical signal itself, minimizing or eliminating optical to electrical or electrical to optical conversion.

The most promising application of integrated optics is in the field of optical communication, optical signal processing and spectrum analysis.

Some important features of the integrated optics are:

1. High speed data transmission, large bandwidth and high sensitivity.
2. Immunity to mutual interference and cross talk.
3. Minimized capacitative and electrical loading delay effects.
4. Easy to control guided wave by electro optic, magneto optic, acousto optic, thermo optic effect.
5. Large optical density allows to utilize non-linear optical phenomena.
6. Low power consumption.
7. Possible to obtain a high density optical components.
8. Good mechanical, thermal and alignment stability.
9. Mass producible with high precision, reproducibility and reliability.
10. Compact and light weight.

11. Low cost.
12. Most devices are based on single mode plannar optical waveguide.

As in any opto electronic device where electrons and photons are involved in producing device characteristics, in an optical IC multiple functions are separately performed by electronic and opto electronic devices. These integrated optic devices are constructed on different thin film dielectrics provided with interconnects on a single waveguide structure. Various methods such as thin film deposition, ion exchange diffusion, photo-lithography, epitaxial growth, selective etching, electron beam writing lithography, proton exchange etc. have been utilized in the fabrication of integrated optical devices.

10.2
WAVEGUIDE STRUCTURE

A waveguide is a dielectric region placed in between dielectrics of smaller refractive index, the light is propagated through the guiding region on the principle of total internal reflection. This guiding region is in the form of very thin film or strip. The raw materials used and its design is such that loss is very small (<0.1 dB). Lithium niobate and gallium arsenide are two popular materials that are used for integrated optic design. In integrated optics, asymmetric thin film waveguide structure is commonly used, which is made with different refractive indices materials above and below the guiding region.

A planar waveguide consists of a thin film guiding region of refractive index n_1 sandwitched in between a substrate of refractive index $n_2 (n_2 < n_1)$ and top cover of refractive index $n_3 (n_3 < n_1)$. In certain cases, the top layer can be air ($n_3 = 1$) which is the configuration with integrated optical components open to air. The wave is allowed to propagate along the z-direction confining along the x-direction only; the light can diffract along the y-z plane.

In three-dimensional waveguides, the guided mode is effectively confined in both x and y directions without spreading due to diffraction effect on the guide surface. Such a lateral confinement can be obtained in a stripe waveguide which consists of a thin film stripe region of high refractive index n_1 surrounded by lower refractive index regions. The direction of propagation is the z-direction with light confinement along both x and y directions.

If the thickness of the guiding region is d, the guide can support a mode of order m, provided

$$d \geq \frac{\lambda [m + (1/2)]}{2(n_1^2 - n_2^2)^{1/2}}$$

where, $m > 0$, λ is the wavelength with in the guide. A variety of other waveguides such as ridge, corner bent, S-shaped, tapered, etc. are also used for different applications.

The waveguide is constructed by creating a refractive index variation in the guiding medium for providing effective light confinement. A wide variety of techniques such as sputtering, liquid phase epitaxial growth, indiffusion, etc. are

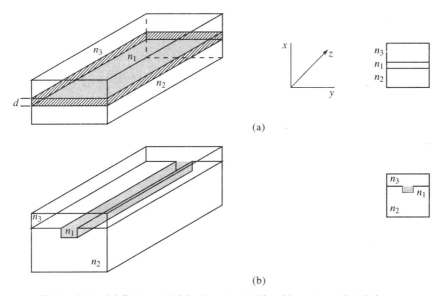

Figure 10.1 (a) Planar and (b) stripe waveguide with cross-sectional views.

employed. Using heterostructures, the refractive index difference regions are made by different bandgap materials. The refractive index lowering can also be achieved by introducing the charge carriers in the guiding region.

10.3
WAVEGUIDE DEVICES

An optical integrated circuit is constructed by incorporating various optical devices in the waveguide structure, each of them perform its characteristic function to get the desired output. Generally, these elements are classified as passive devices and active devices. Passive devices can produce a static change for optical waves such as coupling, bending, focussing, expanding filtering, splitting, isolating, reflecting etc. The components that can produce these functions are implemented by modifying the waveguide structure. The active devices produce a dynamic change to the light, i.e. involve the energy conversion such as light generation and detection, modulation, switching, oscillation, amplification, etc. These devices control the light properties using externally applied physical phenomena.

10.3.1 Passive Waveguide Devices

Passive waveguide devices are constructed by modifying the waveguide structure. Some of these devices are discussed as follows:

Waveguide Lenses

These are used to focus, collimate or expand guided modes in an integrated optical circuit, and are important component for signal processing, imaging, fourier transformation, etc. The lens function based on ray bending can be achieved by changing the mode index (i.e. the effective index of refraction for a guided wave). The mode index increases with the increase in the thickness of the guiding layer or cladding and also with impurity diffusing, which can cause ray deflection according to Snell's law. Depending on the mode index of the lens area convex or concave lens function is obtained. This type of waveguide lens is called mode index lens (Figure 10.2).

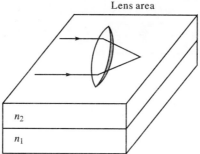

Figure 10.2 Mode index lens.

Luneberg lens which has a rotation—symmetrical refractive index profile and geodesic lens in which a ray travels along a line connecting two points on a curved surface with minimum length are also used in waveguide structure.

Diffraction Lenses

Diffraction lenses are classified into Fresnel zone lenses and grating lenses, which are based on light diffraction in a periodic structure. Fresnel zone lens can be realized with a gradient thickness or index distribution. For a gradient thickness type, thickness distribution is

$$L(x) = L_{max} \left[\frac{\phi(x)}{2\pi} + 1 \right]$$

with Δn = constant and for a gradient index type, the index distribution is given by

$$\Delta n(x) = \Delta n_{max} \left[\frac{\phi(x)}{2\pi} + 1 \right]$$

with L = constant.

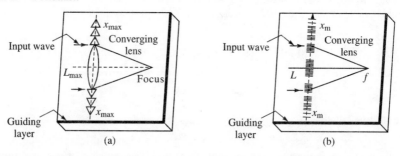

Figure 10.3 (a) Graded thickness Fresnel zone lens, (b) Graded index Fresnel zone lens.

Here, L is the lens thickness, Δn is the difference in mode index between inside and the outside lens area and $\phi(x)$ is the phase modulation of the guided wave propagating along the z-direction.

The lens function in a grating lens is obtained by making a chirped grating with a modulation in the mode index.

$$\Delta n(x) = \Delta n \cos \phi(x)$$

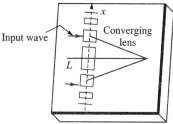

Figure 10.4 Grating lens.

Light Bending Devices

The optical path bending devices are employed to change the path direction or to translate the path of light as required in a waveguide structure. Some of these devices are explained as follows:

1. ***End face mirror*** The end faces at the right angle with respect to the waveguide surface are polished accurately. Additional high reflection coating may also be given to increase the reflection coefficient. The reflection about 100% at the face can be achieved under total internal reflection condition.

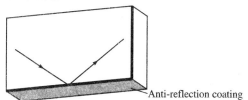

Figure 10.5 End face mirror.

2. ***Prism*** A thin film prism is loaded on a slab waveguide. The beam bends by refraction according to Snell's law at the boundary of two media which have different refractive indices.

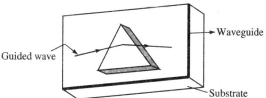

Figure 10.6 Path bending by thin film prism.

3. **Ridge** The guided wave is reflected by a thicker to thinner transition in the waveguide, where the mode index effectively changes from higher to lower.

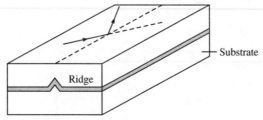

Figure 10.7 Ridge path bending.

4. **Bent waveguides** Corner bent waveguide consists of two or more straight waveguides in the x and y directions with minimum connection loss. An S-shaped waveguide is formed by connecting two waveguide with a uniform radius of curvature.

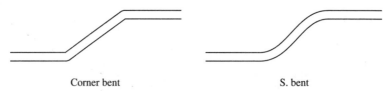

Corner bent   S. bent

Figure 10.8 Bent waveguide for path bending.

5. **Reflection and transmission** Reflection and transmission type grating elements can be incorporated in the waveguide structure as path bending components. They are also called as path bending Bragg gratings.

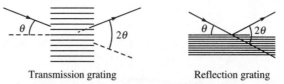

Transmission grating   Reflection grating

Figure 10.9 Path bending by Bragg grating.

Optical Power Dividers

Optical power dividers are used to divide the optical power into two or more branches in a waveguide. A symmetric 2-branch waveguide has a linear tapered region which divide the optical power into two branches (Figure 10.10a) such that the incident fundamental mode propagate smoothly without mode conversion. Near the coupling region dimension, refractive index distribution and branching angle are carefully controlled. The power is monitored by controlling the branching angle. By connecting two branch components multiple power division function can be performed (Figure 10.10b). In a single mode structure, the power division can be obtained at the required ratio, whereas it is difficult to achieve it in multimode structure, because the power dividing ratio depends on the excitation conditions of the guided mode.

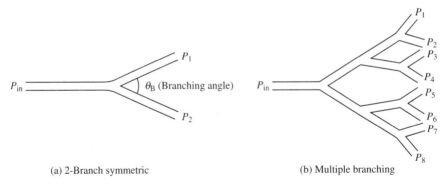

(a) 2-Branch symmetric   (b) Multiple branching

Figure 10.10 Optical power divider.

Directional Couplers

Couplers are devices used to interconnect adjacent waveguides to combine or divide light with very low loss. A directional coupler allows the light to pass through one direction only. In the 1 × 2 directional coupler shown in Figure 10.11

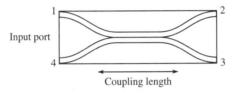

Figure 10.11 Directional coupler 1 × 2.

divide the input power applied at port (1) into two outputs at ports (2) and (3). In ideal condition no light reaches at the port (4). For efficient coupling it requires that the propagation vectors are the same for each waveguide. The light coupled can be varied from 0 to 100 percent by suitably choosing the coupling length—the length over which a complete transfer of optical power is taking place. A star coupler is a $n \times m$ uni/bidirectional multimode coupling device which has many input/output ports as shown in Figure 10.12.

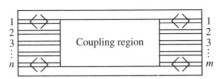

Figure 10.12 $n \times m$ uni/bidirectional coupler.

Various coupling functions that can be incorporated in waveguides are as follows:
1. Combiner: Combiner combines a number of input into one or more outputs.
2. Splitter: It divides input into two or more outputs.
3. Polarizing splitter: In polarizing splitter, the input signal divide into two orthogonal polarized light.
4. Monitor: Couple some portion of the light to the monitor port.

Waveguide Polarizer

Waveguide polarizers are used to remove one of the orthogonal mode i.e. TE or TM mode. For example, a metal cladded waveguide, which is fabricated with a metal (e.g. Al) cladding on the glass waveguide structure, the TM mode is absorbed by the metal film and the TE mode is allowed to transmit as shown in Figure 10.13.

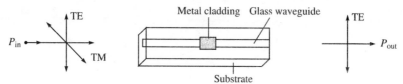

Figure 10.13 Metal cladding waveguide polarizer.

Waveguide polarizer can also be formed by utilizing the double refraction property of anisotropic crystals such as calcite crystal cladded on glass waveguide structure. The refractive indices of the ordinary ray and extraordinary ray in calcite are $n_o^- = 1.656$ and $n_e = 1.458$ respectively, and the effective index of the guided mode in the glass waveguide (n_g) is such that $n_o > n_g > n_e$. Then if the calcite cladding is formed such that its optic axis is parallel to the TE mode, the TE mode can propagate without loss while TM mode which is absorbed by the calcite. (Figure 10.14). Such waveguided polarizer can also be formed with Nb_2O_3 film cladded on $LiNbO_2$ waveguide. This type of waveguide polarizers has a smaller insertion loss as compared with the metal-clad polarizers.

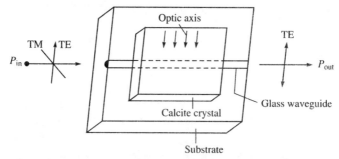

Figure 10.14 Waveguide polarizer using calcite crystal.

Wavelength Multiplexers and Demultiplexers

Wavelength multiplexers and demultiplexers are devices used in a waveguide structure to combine or split a guided wave constituting many wavelength components. Angular dispersive devices such as prisms, gratings, etc. are employed to implement these functions because of their higher dispersion efficiency and the integration capability. These dispersive devices convert the angular dispersion into lateral displacement with the help of lenses. (The angular spread $d\theta$ between two wavelengths separated by $d\lambda$ is determined by the angular dispersion $d\theta/d\lambda$). The basic mechanism is shown in Figure 10.15. They can be constructed by prism, reflection or transmission gratings arrays with different grating elements or orientation, chirped gratings, etc.

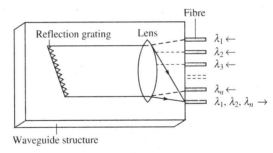

(a) Wavelength multiplexer

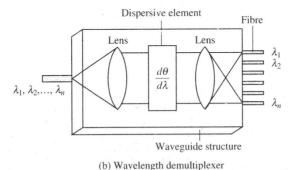

(b) Wavelength demultiplexer

Figure 10.15 Wavelength multiplexers and demultiplexers.

10.3.2 Active Waveguide Devices

Active waveguide devices are used to control the guided wave effectively in a waveguide structure by utilizing some externally applied physical phenomena such as electro optic, acousto optic, magneto optic, thermo optic or non-linear optic effects. The advantages of these active waveguide devices over the conventional bulk devices are as follows:

1. Light spreading due to diffraction can be eliminated by light confinement in the waveguide.
2. By monitoring the externally applied effect, the guided mode can be controlled effectively, since the interaction between light and the external effect is restricted only in the region surrounding the waveguide.
3. Compactness and very small size of the electrodes considerably reduces the drive voltage.
4. Many desired functions can be realized by using waveguide devices.

Active Waveguide Devices based on Electro Optic Effect

A large number of active devices utilizing electro optic effect can be formed in a waveguide structure with an appropriate design of electrodes on the waveguide. Lithium niobate ($LiNbO_3$) is an electro optic crystal which can be used to fabricate low loss single mode waveguide by doping with Ti atoms. They are divided into following types according to their operations:

1. *Phase control type*   Here the output light is phase modulated by applying voltage to the planar electrodes on a straight waveguide, e.g. Phase modulator, Mach-Zehnder interferometric modulator.
2. *Directional coupler type*   Here the externally applied voltage causes a change in the distributed coupling between the adjacent waveguides due to the phase modulation of the guided mode. This results into light intensity modulation at the output end, e.g. wavelength filter, modulator switches.
3. *Index distribution type*   Optical path can be changed by an applied voltage with appropriate electrodes by controlling the refractive index distribution inside a branching waveguide, e.g. total internal reflection switch and cut-off switch.
4. *Electro optic grating type*   Deflection and mode conversion of guided modes are possible by using periodic index change inside the waveguide with interdigital electrodes, e.g. light deflector and TE-TM mode convertor.

10.4

PHASE MODULATOR

An electric field applied to an electro optic material will change the refractive index of the medium by virtue of Pockel effect. This change in refractive index then Pockel utilized to modulate the phase of the propagating mode. The phase modulator shown in Figure 10.16 consists of a waveguide strip fabricated on lithium niobate substrate by indiffusion of titanium atoms (which increases the refractive index).

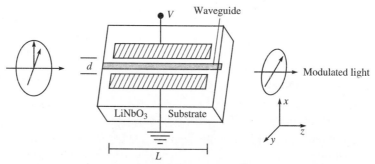

Figure 10.16   Waveguide phase modulator

On either side of the waveguide channel, two coplanar film electrodes (with Al) are evaporated in order to apply the modulating voltage. The applied field is along the y-direction and predominantly polarized along the y-directions. Note that the induced electric field is not uniform between the electrodes and also all the applied field lines are not using inside the waveguide. The light is propagating along the z-direction which can be resolved into two orthogonal modes (E_x and E_y). The external voltage applied to the electrodes induces a change in refractive index so that the phase introduced between E_x and E_y is

$$\Delta\phi = \Gamma \frac{2\pi}{\lambda} n_e^2 r \frac{L}{d} V$$

where

$\Gamma = 0.5$ to 0.7 which depends on the asymmetry of the field in between the electrodes

n_e = Effective refractive index
r = Electro optic coefficient
L = Modulator width
d = Spacing between the electrodes
V = Modulating voltage
λ = Wavelength of the light

10.5

MACH–ZEHNDER INTERFEROMETRIC MODULATOR

A waveguide Mach–Zehnder interferometric modulator is formed by implanting a single mode stripe waveguide structure as shown in Figure 10.17 on a LiNbO$_3$ substrate (or other EO materials). The input at A divides into two branches (1 and 2) and rejoin at B to form output. Splitting and combining at A and B involve two y-junction single mode waveguides. Light entering through the input at A split into two branches and interfere at B, the output port. The electric fields are applied via planar electrodes on either arms of the waveguide channel. Note that the electric field applied is in opposite direction in two branches. The electric field induces a refractive index change inturn which produces a phase change (in the opposite direction) in propagating modes in two waveguide branches. The applied voltage controls the phase difference between the two interfering waves at B which leads to the modulation of the output light intensity.

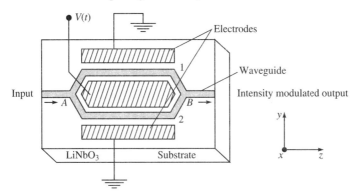

Figure 10.17 Waveguide Mach–Zehnder interferometric modulator.

10.6

DIRECTIONAL COUPLER SWITCH

As we studied in the section of directional coupler, in a single mode waveguide directional coupler two waveguide structures are arranged very close to each other, so that the evanescent fields will overlap and there is an energy exchange between the two waveguides. There is a complete transmission of light from waveguide A to B when the propagation constant of the two modes are identical (i.e. when $\beta_A = \beta_B$ the phase mismatch $\Delta\beta = 0$). Whereas there is only partial transfer of energy between the two waveguides when the propagation constants are not equal (i.e. $\beta_A \ne \beta_B$, the phase mismatch $\Delta\beta \ne 0$). The propagation constant of the modes in two waveguides can be altered by applying an electric field via electro optic effect. In the presence of applied electric field since the refractive index changes oppositely, there is a net refractive index change of $2\Delta n$, so that the phase mismatch is

$$\Delta\beta = 2\Delta n \frac{2\pi}{\lambda}$$

$$= 2\left(\frac{1}{2}n^3 r \frac{V}{d}\right)\frac{2\pi}{\lambda}$$

The state with light entering from waveguide A and emerging from B is referred to as the cross state, while the state with light entering in waveguide A and emerging from the same waveguide is called the parallel state. For going over from the cross state to the parallel state, the phase match required to be

$$\Delta\beta = \frac{\sqrt{3}\pi}{L}$$

where L is the transmission length corresponding switching voltage is given by

$$V_0 = \frac{\sqrt{3}\lambda d}{2n^3 rL}$$

V_0 depends on the refractive index, electro optic coefficient and also the geometry of the waveguide (L). (The coupling efficiency is also related to L). By applying a voltage, light can be switched between the waveguide from one to other, and the device is known as integrated directional coupler switch.

An integrated directional coupler switch is fabricated by implanting two single mode stripe waveguide on $LiNbO_3$ substrate. The waveguides are very close to each other over a length L. Two electrodes are also deposited over the waveguide (as shown in Figure 10.18) over the interaction region. The electric field points in the opposite direction thus creates an opposite refractive index change in the two waveguide. The switching can be achieved by setting the applied voltage.

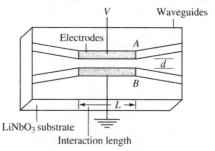

Figure 10.18 Directional coupler switch.

10.7
ACTIVE WAVEGUIDE DEVICES BASED ON ACOUSTO OPTIC EFFECT

The change in refractive index of a medium due to mechanical strain produced by an acoustic wave is called the acousto optic effect. The periodic variation of the index of refraction with respect to the applied acoustic waves lead to form a refractive index grating. When light falls on such a refractive index grating, Raman–Nath or Bragg type diffraction will take place. Utilizing this effect many active waveguide devices can be implemented. Some of them are modulators, switches, mode converters, wavelength filters, deflectors, spectrum analyzers etc.

These devices utilizing surface Acoustic Waves (SAW) have following advantages:

1. The SAW wavelength can be varied over a wide range by varying the frequency of the electrical signal. This leads refractive index grating with variable periods.
2. The time dependent input signal can be converted to a spatially distributed output. This is achieved by spatial modulation of the refractive index grating.

However, acousto optic response time is limited due to the smaller velocity of acoustic wave in the medium as compared with the electro optic devices.

10.7.1 Acousto Optic Bragg Modulator

Bragg diffraction is widely utilized due to its high diffraction efficiency. An acousto optic Bragg modulator shown in Figure 10.19 consists of a y-cut LiNbO$_3$ substrate with Ti indiffused single mode thin film waveguide. A Surface Acoustic Wave (SAW) is formed with uniform period interdigital electrode finger transducer. The SAW power is the RF power that is fed to the transducer. All the RF transducer power can be transferred to the SAW by matching the impedances of the RF source and the interdigital transducer. The SAW propagates along the z-axis of the LiNbO$_3$ and diffracted guided wave propagates along the x-axis.

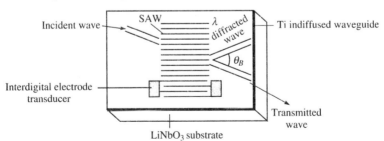

Figure 10.19 Acousto optic Bragg modulator.

The deflection angle is given by the Bragg angle

$$2\theta_B = 2\sin^{-1}\frac{\lambda}{2n\Lambda} \approx \frac{\lambda}{n\Lambda}$$

where λ is the wavelength and n is the effective index of the optical guided mode. The wavelength of the SAW is given by

$$\Lambda = \frac{V}{f}$$

where V is the velocity and f is the frequency of the acoustic wave. The diffraction efficiency and the diffraction angle are controlled by the SAW power and frequency respectively. The response speed is determined by the SAW transit time, i.e.

$$\tau = \frac{\text{Interaction region length}}{\text{Velocity of the SAW}}$$

10.7.2 Acousto Optic Spectrum Analyzer

Figure 10.20 shows an integrated acousto optic spectrum analyzer which is formed by a plane optical waveguide on the surface of a suitable substrate such as LiNbO$_3$.

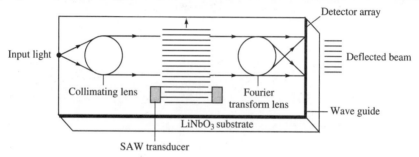

Figure 10.20 Acousto optic spectrum analyzer.

Collimated light from a waveguide collimating lens is allowed interact with the SAW. The SAW is produced by an interdigital transducer. The angle at which the beam gets diffracted is proportional to the SAW frequency so that if the interdigital transducer is excited simultaneously with different frequencies, the Bragg diffracted light waves will propagate along different directions. These beams are then focussed on the plane of a detector array using a waveguide fourier transform lens. This device functions as a spectrum analyzer which gives a real time spectrum of a wide band signal.

10.8
ACTIVE WAVEGUIDE DEVICES BASED ON MAGNETO OPTIC EFFECT

Magneto optic waveguide devices are implemented by using magneto optic material for the guiding medium. A light wave propagating through the waveguide, in the presence of an applied magnetic field induces a large Faraday effect at very low driving power. The rotation of polarization is proportional to the propagation length and the propagation constants are different for TE and TM modes in the waveguide. The most important application is the implementation of devices such as optical isolators.

Figure 10.21 shows a magneto optic isolator using a semi leaky type waveguide. Here the waveguide is cladded with LiNbO$_3$ slab, with the crystal axis tilted in the waveguide plane. The forward TE wave guided, while the backward wave converted into TM wave and leaks into LiNbO$_3$ cladding.

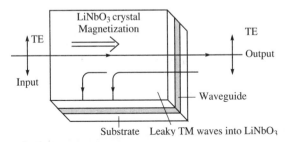

Figure 10.21 Magneto optic isolator waveguide.

10.9
ACTIVE WAVEGUIDE DEVICES BASED ON THERMO OPTIC EFFECT

Change in refractive index induced by temperature is termed as thermo optic effect. In thin film waveguides thermo optic modulation and switching with response time of the order of milliseconds to microseconds is possible. They require that temperature much higher than the operating temperature of the device. K$^+$ ion exchanged glass waveguide is an important material for the implementation of thermo optic based waveguide devices. Figure 10.22 shows a thermo optic branching waveguide switch formed by K$^+$ ion exchanged glass waveguide. The waveguide at the output smoothly expanded and divided into two branches. In the vicinity of the branching region two identical titanium sputtered film heaters of resistance about 1.8 kΩ are placed along with SiO$_2$ buffer layer. In the absence of any applied voltage to the heater the input light power is equally divided into two branches. When the voltage is applied, the thermally induced refractive index lead to a control over the lateral confinement of the guided mode. For example, when the voltage is applied to the heater A the guided modes are confined in the higher refractive index region so that more power reaches to the branch A, while less power to the branch B.

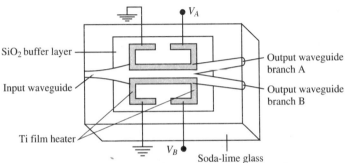

Figure 10.22 Thermo optic branching waveguide switch.

10.10
ACTIVE WAVEGUIDE DEVICES BASED ON OPTICAL NON-LINEARITY (OPTICAL BISTABLE DEVICE)

Optical bistability is a non-linear phenomenon which has two stable states of transmission. This bistable state can be observed by controlling the transmission by means of positive feedback from the optical output. For example, an electro optic controlled element is inserted in a Fabri perot cavity, and the transmitted power is converted to voltage through an optical detector. This signal is then fed back to the electro optic controlled element to develop bistable characteristics.

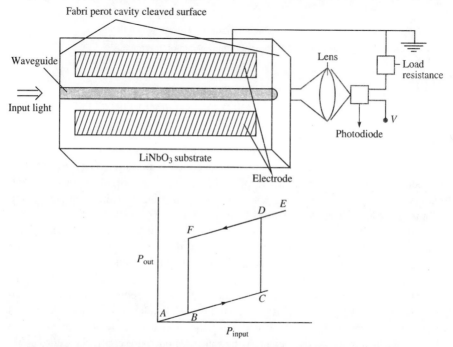

Figure 10.23 Waveguide optical bistability device and its input-output characteristics.

10.11
WAVEGUIDE COUPLING

Coupling of light with waveguide is very important for efficient performance of the waveguide devices. Several techniques are employed for this purpose. Some of them are given as follows:

Direct Edge Coupling

Edge of the waveguide film is directly attached to the source/detector/device [Figure 10.24(a)].

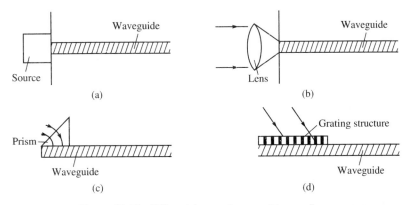

Figure 10.24 Different types of waveguide couplings.

Edge Coupling with Lens

A lens is used to comple the collimated light in the waveguide [Figure 10.24(b)].

Prism Coupling

Incident light beam enters the prism under goes critical angle reflection and propagates [Figure 10.24(c)].

Grating Coupling

A grating diffracts an incident beam into one or more transmitted waves; and coupling occurs if the longitudinal propagation factor is equal to that of the propagating mode [Figure 10.24(d)].

10.12
OPTO ELECTRONIC INTEGRATED CIRCUIT

Opto electronic integrated circuit involves the monolithic integration of electronic and optical devices and their interconnects on a same semiconductor chip. The integration involves all the active and passive devices such as sources, detectors, optical fibres, amplifiers, couples, filters deflections, modulators, switches and so on. The performance such an optoelectronic integrated circuit (also known as optical IC) depends on the functioning of the individual electronic and optical devices used, and the coupling between them. Since hetero structures and different fabrication processing techniques are involved for each component, the realization of such devices poses many challenges. They have the advantages of high speed, high sensitivity, compactness, reliability, easy in connection into a larger network and low cost. Opto electronic integrated circuit devices are successfully fabricated for achieving following applications:

1. Optical switch
2. A.D. converter
3. Optical fibre interconnect

4. Signal processing devices (correlators etc)
5. Sensors (Temperature, pressure)
6. Spectrum analyzer.

REVIEW QUESTIONS

1. What are the features of integrated optics?
2. Mention some applications of integrated optics.
3. Explain the waveguide structure used in integrated optics device fabrication.
4. What is meant by active and passive waveguide devices? Give some examples.
5. Explain design of following waveguide devices.
   (i) Lens
   (ii) Light sending devices
   (iii) Power dividers
   (iv) Directional couplers
   (v) Polarizers
   (vi) Multiplexers and demultiplexers
6. Explain active waveguide devices based on EO, AO, MO, TO, NLO effect with example.
7. Explain methods of waveguide coupling.
8. What is an opto electro integrated circuit?

11

Holography

11.1
INTRODUCTION

Holography is a method of recording optical images in three dimensions. Conventional photograph is a two-dimensional recording of an object using a non-coherent light source, the image of which represents the intensity distribution prevailed on the photographic film. In this case, since the emulsion on the photographic plate is sensitive only to the intensity variation (i.e. the square of the amplitude) the phase information of the light is lost while recording.

The method of holography, also known as wavefront reconstruction, was first introduced by Dennis Gabor in 1949, by recording not only the amplitude but also the phase of the wave using a mercury lamp as the light source. The word holography is originated from Greek word *holo* meaning whole and *graphein* meaning recording. Thus, holography implies complete recording. Wide popularity is gained to holography due to its capability to produce 3D images as true as the objects. Light with a high degree of coherence is required for the realization of holography hence the technique became popular with the advent of lasers.

A hologram is a process of recording both the intensity and relative phase of the light wave at each point of the interference pattern formed between two coherent light beams which are coming from the same source on a light sensitive medium such as a photographic plate. The interference pattern recorded in the hologram bears no resemblance to the original object at all, but possesses all the information of the object in a coded form that is needed to reconstruct the three-dimensional image.

11.2
METHOD OF HOLOGRAPHY

The principle of holography is based on the interference of two highly coherent light beams such as laser and consists of following two steps:

1. Recording of hologram
2. Reconstructing of image

11.2.1 Recording of Hologram

An expanded laser beam is divided into two by a beam splitter. One portion of the beam is directed to illuminate the object whose hologram is to be recorded. Each point on the object scatters light and acts as a point source of spherical wave. This wave is known as the object wave and is allowed to fall on the photographic plate. The other portion of the beam called the reference wave which is a plane wave is allowed to fall directly on the photographic plate. (Figure 11.1). The superposition of these two beams (object wave and reference wave) produces an interference pattern of both the amplitude and relative phase on the plane of the photographic plate and are thereby recorded. The developed negative of the interference fringe pattern is known as hologram. The interference pattern is a complex pattern of lines and swirls with a very fine spacing as small as about μm. The image embedded in it bears no resemblance to the object at all, but contains all the information about the geometrical characteristics of the object in an encoded form.

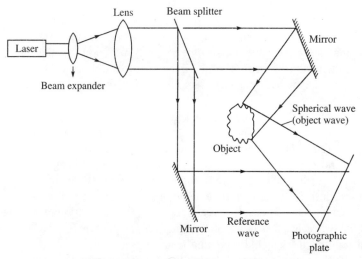

Figure 11.1 Recording of a hologram.

11.2.2 Reconstructing of Image

The method of reproducing the image embedded in the hologram is known as reconstruction. For reconstructing, the hologram is illuminated by a parallel beam from a laser source. This illumination beam called the reconstruction wave, is identical to that used during the formation of the hologram. Most of the light passes through the complex pattern of the hologram which is acting as a diffraction grating.

The process, in general, form two images—a virtual image behind the hologram and the real image on the eye side (see Figure 11.2). The reconstructed

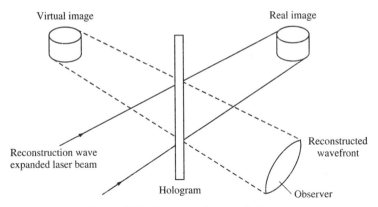

Figure 11.2 Reconstruction of a hologram.

wavefront appears to be coming from the object, and the virtual image has a complete three-dimensional form. The perspective of the image will change as eyes move from the viewing position.

The method described in the previous paragraph is known as off axis or carrier wave holography which was introduced by Leith and Upatnicks in 1964.

11.3
CONDITIONS FOR RECORDING HOLOGRAMS

Some of the conditions to be satisfied for recording good hologram are mentioned as follows:

1. As holography is an interference based technique, a light source with a high degree of coherence (both spatial and temporal coherence) should be employed to obtain high contrast fringes.
2. The state of polarization of the reference and object wave should be nearly the same. Usually a laser source with vertical electric field vector is employed.
3. The holographic set up should be arranged such that the vibrations due to mechcnical, acoustic or thermal effects should not produce an optical path variation more than $\lambda/8$.
4. An optical attenuator is usually inserted in the path of the reference beam to reduce the irradiance 1/3 to 1/15 times.
5. The fine structure of the interference fringes requires special photographic emulsion with a high resolving power.

11.4
HOLOGRAM RECORDING MATERIALS

Holography requires materials of high resolution capabilities. Photoemulsion is the most commonly used sensitive medium. The most significant phenomenon affecting hologram formation is the relationship between the developed film transmittance and the incident light intensity. Other materials used for holographic recording are thermoplastics, photochromics, photo resist, dichromated gelatin, electro optic crystal, magnetic films etc. Recording processes involve the following

1. Reduction to Ag metal grain bleached to silver salts
2. Photo cross linking
3. Photopolymerization
4. Electrostatic imaging
5. Photoinduced absorption bands
6. Electro optic effects

11.5
PRINCIPLE OF HOLOGRAPHY

Let us assume that the electric field of the reference wave has the form

$$E_r \approx e^{i\alpha x} \tag{11.1}$$

and that of the object wave is represented by $E_0(x, y)$. The amplitude and phase of the object wave are included in its representation. Both waves are assumed to be in the same state of polarization. They interfere in the plane of the photographic plate, the recorded irradiance is given by

$$I(x, y) = |E_r + E_o|^2 \tag{11.2}$$

$$= (E_r + E_o)(E_r^* + E_o^*)$$

$$= |E_r|^2 + |E_o|^2 + E_o E_r^* + E_r E_o^* \tag{11.3}$$

Substituting the value of E_r from Eq. (11.1) in Eq. (11.3), we get

$$I(x, y) = 1 + |E_o|^2 + E_o e^{-i\alpha x} + E_o^* e^{i\alpha x} \tag{11.4}$$

The first two terms represent the background irradiance and the other two terms represents the interference pattern due to reference and the scattered (object) wave. The photographic plate is then processed to form a hologram.

By suitable developing processes the transmittivity $\tau(x, y)$ of the hologram can be made related linearly to irradiance $I(x, y)$ such that

$$\tau(x, y) = \tau_o - Bt I(x, y) \tag{11.5}$$

where τ_o is the transmittivity without exposure, t is the exposure time and B is a constant. To reconstruct the object, hologram is illuminated by the reconstruction wave, which is same as the reference beam. The transmitted light field is then given by

$$E_t^{(xy)} = \tau(x, y) e^{i\alpha x}$$

$$= [\tau_o - Bt I(x, y)] e^{i\alpha x}$$

$$= [\tau_o - Bt(1 + |E_o|^2 + E_o^* e^{i\alpha x} + E_o e^{-i\alpha x})] \times e^{i\alpha x}$$

$$= (\tau_o - Bt) e^{i\alpha x} - Bt |E_o|^2 e^{i\alpha x} - BtE_o(x, y) - BtE_o^*(x, y) e^{2i\alpha x}$$

$$= [\tau_o - Bt(1 + |E_o|^2)] e^{i\alpha x} - BtE_o(x, y) - BtE_o^*(x, y) e^{2i\alpha x} \tag{11.6}$$

The hologram acts like a diffraction grating. The first term represents the direct beam, the second term represents the virtual image field, which is proportional to the object wave field and the last term form the conjugate (real) image which is phase modulated by a factor exp($2i\alpha x$).

11.6
TYPES OF HOLOGRAMS

11.6.1 Volume Holograms

A volume hologram is essentially a thick grating in which the whole volume of the recording medium takes part in the hologram construction process i.e. the irradiance distributions exhibits inside the volume of the recording medium. There is no sharp separation between a thin and a thick hologram, but generally a hologram with

$$\frac{2\pi\lambda_m T}{d^2} > 10$$

is considered as a volume hologram. Here λ_m is the wavelength in the medium of thickness d, and T is the exposure time.

A volume reflection hologram can be constructed by directing object wave (scattered from the object) and the reference wave from opposite direction (which form a standing wave) on the thick recording medium. The resulting interference fringes generated by the standing waves are recorded in the depth of the thick emulsion. The hologram can be reconstructed by illuminating white light, and hence, they are also known as white light holograms. A volume hologram is highly selective to wavelength and misorientation.

11.6.2 Rainbow Holography

Rainbow holography is a technique for producing hologram images using white light. Here the hologram is constructed with white light and the images are observed through the transmitted light and are seen in the rainbow colours. This type of hologram is capable of producing brighter and more colourful holograms.

11.6.3 Colour Holography

Colour holography is a technique of producing three-dimensional multicoloured images of an object with good colour reproduction. To produce multicoloured hologram, the object is illuminated with three primary coloured coherent beams. The scattered light from the object and the reference beams are then allowed to fall on the recording medium so as to produce reflection type hologram. Interference fringes are formed throughout the depth of the emulsion. The hologram may be considered as an incoherent superposition of three primary holograms. Reconstruction of the hologram will generate nine virtual and nine conjugate images. Three primary images are exactly superposed to give multicoloured recombination and the other six images called the cross-talk images are eliminated. The coherent light sources commonly used for three primary colours (blue, green and red) of light are of the following types:

1. 488 nm Argon ion laser
   532 nm Second harmonic of Nd–Yag laser
   633 nm Helium–Neon laser
2. 488 nm (Argon ion laser) or 442 nm Helium Cadmium laser
   514 nm (Argon ion laser)
   633 nm (Helium–Neon laser)
3. 476 nm Krypton ion laser
   521 nm Krypton ion laser
   647 nm Krypton ion laser

One of the major difficulty in colour holography has been the elimination of cross talk images.

11.7
CHARACTERISTICS OF A HOLOGRAM

1. Holography is a method of recording images in three dimensions in which both amplitude and phase are recorded in the form of interference pattern. The virtual image produced on reconstruction of the hologram produces three-dimensional images which is looking same as that of the real image.
2. The information recorded by holography is in the form of interference pattern and does not produce negative of the image as in the case of an ordinary photography.
3. The pattern recorded on the hologram has no resemblance to the object at all, i.e. the information is embedded in a coded form.
4. While recording, each point on the object scatters light over the whole area of the hologram, and therefore, each point on the hologram contains information about the whole object. Each part of the hologram is capable of reproducing the whole image of the object. Hence, even if a hologram is distructed into pieces or a part of the hologram is broken, it will not destroy the specific portion of the image (as in ordinary photograph). However, as the size of the hologram decreases, the image resolution will decrease.
5. Holograms have extremely high information storage capacity. A large number of scenes can be independently stored on a hologram simply by rotating the plate while recording, and can be reconstructed each scene by properly orienting the hologram with respect to the readout wave.
6. White light hologram, coloured hologram moving image holograms, etc. can be developed.

11.8
APPLICATIONS OF HOLOGRAPHY

Holography finds many scientific, technological, commercial and entertainment applications. In many applications three dimensional characteristics and high data storage capacity of the hologram is greatly utilized. Some of them are briefly discussed as follows:

Holographic Microscopy

In holographic microscopy, the magnified images of a transient microscopic object are recorded. When microscopic particles are in motion, it is difficult to locate and observe all the particles in a volume at a given instant of time. The hologram of the moving particle can be recorded by illuminating the volume using a pulsed laser, so that scene is frozen into the hologram. It can be reconstructed at a later time and is possible to measure the particle size, distribution, the cross-sectional geometry, etc. of the objects. The three-dimensional observation of a transient event with high magnification is the main advantage of the holographic microscopy. The method has been used in droplets, particle size analyzers in cloud chamber, aerosol study, rocket engine exhaust, etc.

Holographic Memory

A holographic computer memory records and reads a large number of bits simultaneously by utilizing holographic technique. Some of the advantages of holographic memory are as follows:

1. Very high storage capacity up to 10^{10} bits/mm^3.
2. As the information is stored over the entire hologram, it is insensitive to small scratches or particle dusts, etc.
3. A large number of bits are read simultaneously allowing a very high readout rate.
4. The recording and reconstruction is insensitive to the exact position of the reference or readout beam.
5. They are easier to align and less subjected to vibrational problems.
6. Several patterns can be recorded on a single hologram.

At present, availability of suitable storage media is the main limitation to the development of such holographic memories.

Holographic Interferometry

Different types of interferometric techniques are used in holography. They are as follows:

1. *Real time holographic interferometry.* Here the interference fringes are viewed in real time.
2. *Double explosure holographic interferometry.* Two sets of object waves recorded one with the object and the other with the object with deformation

by applying a stress on the same plate. Even a very small distortion of the object can be measured by counting the interference fringes.

3. *Time averaged holographic interferometry.* In this case, the object is continuously moving during the exposure. The hologram can be considered as large numbers of exposures at the different position of the surface.

Non-distructive testing is the important use of these interferometric techniques, and are used for determining cracks, weld defects, voids, tyre testing, distortions, vibrations analysis, and so on.

Some of the other applications are: holographic optical elements such as grating, holographic lenses, 3D imaging application, holographic jewellery, display, advertising labels, entertainment uses like cinema, storing confidential data, character recognition etc.

REVIEW QUESTIONS

1. Explain holography. How is it different from ordinary photography?
2. Explain the method of holography.
3. What are the conditions to be satisfied for recording holograms?
4. Mention some hologram recording materials.
5. Explain the principle of holography.
6. What are volume holograms?
7. What are white light holograms?
8. What is rainbow holography?
9. Explain colour holography.
10. What are the characteristic of a hologram?
11. Mention some applications of holography.

12

Display Devices

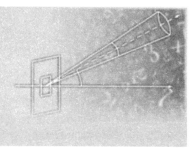

12.1
INTRODUCTION

Display devices play an important role in scientific and technological applications and also in our everyday life which provide (or display) information in the visual form. The technology of display devices is rapidly evolving one. It begins with the old Cathod Ray Tube (CRT) to modern devices such as Liquid Crystal Display (LCD), Light Emitting Diode (LED) display etc. Display systems are broadly categorized into active and passive devices. While active devices emit their own radiation, in passive devices the incident radiation is modulated to display the information.

12.2
CATHODE RAY TUBE (CRT)

A Cathode Ray Tube (CRT) is a device which converts electrical signal into a visual picture form. It has many applications such as in cathode ray oscilloscope, television, computer monitor, and so on. The basic mechanism of operation is that an electron beam generated by thermionic emission is focussed on a fluorescent screen which produces a spot of light wherever it strikes. The essential parts of a CRT are as follows:

Electron Gun Assembly

The electron beam is generated by thermionic emission from a heated cathode coated with oxides of barium, strontium etc. It also consists of control grid to control the electron flow, focussing anode and accelerating anode.

Deflecting Plates

These plates are provided with vertical and horizontal deflecting plates, by which electron beam can be deflected on the screen by applying suitable voltage.

Screen

The screen is made with suitable fluorescent material, which can produce light when an electron beam strikes on it.

Glass Envelope

It is a highly evacuated glass envelope in which the whole arrangements are kept.

Figure 12.1 shows a schematic diagram of a CRT. When an electron beam from the gun assembly strikes the screen, the radiation is generated by cathodoluminescence. The beam deflection is achieved by applying electric fields acting at right angles to the beam direction. The screen is made up of a thin layer of phosphor material granules with a thin layer of aluminium coating. The aluminium layer prevents charge built-up in the phosphor granules and also helps to reflect light towards the observer. By varying the electron beam current, light irradiance can be altered. The phosphor material used should have shorter luminescent decay time than the picture cycle time.

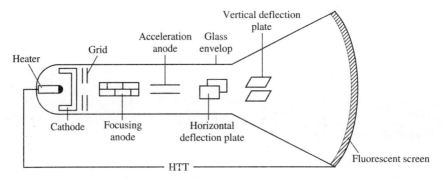

Figure 12.1 Schematic diagram of a CRT.

In normal display operations, the beam is scanned line by line over the viewing area, when one line scan is completed the beam is rapidly switched to the start of the line below. In video applications, the display system consists of 625 lines in Europe (and 525 in America). The entire picture is scanned in 40 ms (1/30 s in America) using raster scan method in which the picture splits up into two interlaced layers. For colour display three electron guns slightly inclined to each other are used (Figure 12.2). A shadow mask which is a metal screen with holes is placed in front of the fluorescent screen. The electron beam coincides at the plane of the shadow mask, this diverges into three electron beam as they pass through the shadow mask holes and produce three light spots on the screen. Each of the three dots emits one of the primary colours—blue, green or red, so that any desired colour can be generated by verifying the relative excitation intensity. Commonly used phosphor are zinc sulfide with silver (blue), zinc cadmium sulfide doped with copper (green) and yitrium oxysulfide doped with europium and terbium (red).

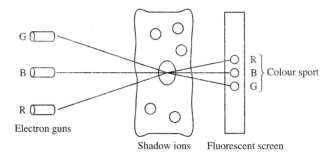

Figure 12.2 Principle of colour display in CRT.

12.3
LIQUID CRYSTALS

Liquid crystal state is a mesophase of matter, a state existing in between solid and liquid over a restricted range of temperature. A rod like shape of molecules (of length about 20–100 Å and width about 5 Å) is a major characteristics of a liquid crystal compound. At high temperatures, it behaves as a liquid with a completely disordered arrangement of molecules, whereas at low temperatures, it shows a completely ordered arrangement of molecules as in crystalline solid. Liquid crystals have low viscosity and can flow (i.e. shows the property of liquids). They also have anisotropic properties similar to that of crystalline solids and also have many optical properties. Their ability to modulate light by applying electric field have made them an important element in the manufacture of display devices.

A large number of organic molecules exhibit liquid crystal state. For example, organic compound like p-azoxyanisole (PAA); 4-methoxy benzylidene 4-butylanaline (MBBA) which has a building block with chemical formula (Figure 12.3) shows liquid crystal state.

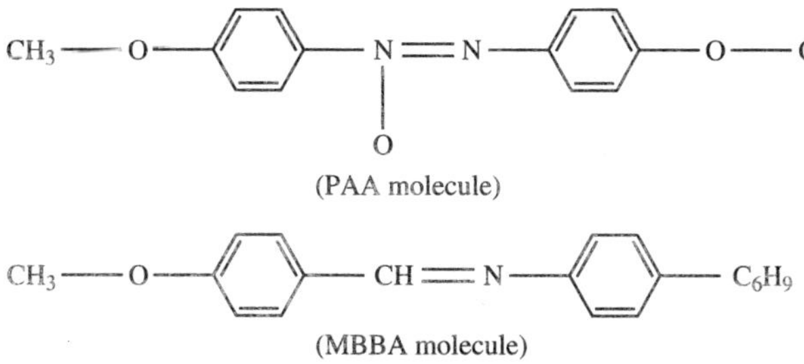

$$CH_3 - O - \langle \rangle - N = N - \langle \rangle - O - CH_3$$
$$|$$
$$O$$
(PAA molecule)

$$CH_3 - O - \langle \rangle - CH = N - \langle \rangle - C_6H_9$$
(MBBA molecule)

Figure 12.3 Building block of PAA and MBBA.

These rod like molecules can take up certain orientation relative to each other in the liquid crystal state. This orientation is described in terms of a quantity called as director, a unit vector pointing along the time averaged orientation of the molecules in a small volume. Depending on their ordering there are three types of liquid crystals which are classified as nematic, cholesteric and smectic structures.

170 Photonics: An Introduction

In nematic structure, rod like molecules are aligned such that the directors line up parallel to each other (Figure 12.4). Even though a long range orientational order is maintained, there is no well-defined positional order preserved. The molecules are free to move relative to each other, hence the phase has liquid property.

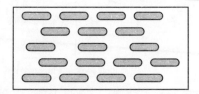

Figure 12.4 Nematic structure of liquid crystal.

In cholesteric structure, orientation of the rod like molecules is maintained as in nematic liquid crystal structure within the layers. However, each layer shows a progressive change in the direction of directors as one moves from one layer to another such that directors between the planes display a helical twist in the material (Figure 12.5). The distance between planes having the same director direction is called the pitch. The pitch is temperature dependent. When a white light is reflected by a cholesteric liquid crystal, it will appear coloured due to Bragg diffraction.

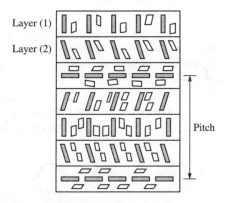

Figure 12.5 Cholesteric structure of liquid crystal.

In the smectic liquid crystal structure, both orientational and positional orientation of the rod like molecules are maintained over a long range. Thus, in a given layer, the rods are arranged in the same direction, and at the same time, the molecules in different layers are also ordered (Figure 12.6).

Figure 12.6 Smectic structure of liquid crystal.

The optical properties of liquid crystals depend on the orientation and twist of rod like molecules in the layers which can be modified by an electric field to produce efficient display devices. Most of the liquid crystal displays are of twisted nematic type.

Another important property due to the interaction of nematic liquid crystal with a solid surface is that the orientation of the rod like molecules can be made to align in a pre-chosen direction. When the liquid crystal materials come in contact with a solid surface which is suitably rubbed with a soft fabric like cotton, the rod like molecules will align along the direction of rubbing. Depending on the direction of rubbing, a twist of 90° can be produced to the rod like molecules as it goes from one surface to the another. Such liquid crystals are known as twisted nematic crystals (Figure 12.7). In homogeneous ordering, the rod like molecules are arranged parallel to the solid surface, whereas in homotropic ordering they are aligned perpendicular to the surface.

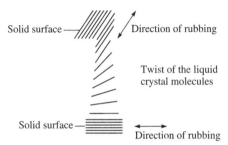

Figure 12.7 Twist of liquid crystal molecules to the interaction with a rubbed solid surface.

By adding cholesteric material to the nematic material, a rotation of 270° to the directors can be achieved. Such materials are known as supertwisted liquid crystals.

Another important property of the liquid crystals is that the direction of ordering of molecules can be altered by applying an external electric field suitably. The amount of light passing through is a function of the applied voltage.

12.4
LIQUID CRYSTAL DISPLAY (LCD)

A simple way of producing liquid crystal display using nematic type liquid crystals is shown in Figure 12.8. The nematic liquid crystals are placed in between two glass plates. The inner surface of the glass plates scratched (being perpendicular to each other on the glass plates) in nanoscale so that the molecular alignment direction of liquid crystals at the walls are at right angle. This causes molecules to undergo a 90° rotation across the cell (Figure 12.7). The cell is arranged in between two polaroids, the polarization direction of each is parallel to the molecular ordering of particular cell surface.

When light passes through the cell the twisted nematic structure will rotate the polarization direction through 90° as the light propagate. The vertically polarized light entering through the cell will become horizontally polarized at the other

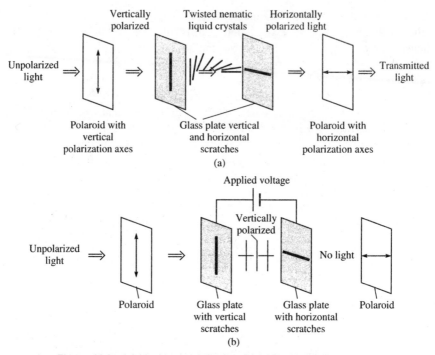

Figure 12.8 (a) Liquid crystal display, (b) LCD with applied voltage.

side (Figure 12.8a). When an electric field is applied, the direction of polarization within the cell is such that it is not rotated, and hence, cannot pass to the other side. Thus, the cell can be appeared bright or dark depending on the voltage level applied to each pixel of the display. A complete range of colours can also be achieved by placing a mask containing an array of red, green and blue filters arranged in groups of the pixel and by controlling the brightness. Liquid crystal displays have a relatively slow response and the angle of viewing is restricted.

12.5
CHARGE COUPLED DEVICES (CCD)

Charge coupled device is a 2 × 2 array of small optical detectors used to image a spatially varying information of an object. Each detector is called a pixel which detects the light at the corresponding part of the wavefront. The centre to centre distance between two adjacent pixels ranges from 10 to 40 μm, and usually the arrays of 256 × 256 to 4096 × 4096 pixels are available.

The basic structure of a CCD is a shift register formed by an array of closely spaced potential well capacitors. It consists of a thin layer of SiO_2 grows on a silicon substrate with a transparent electrode coating at the top. This acts as a tiny capacitor. When a positive potential is applied to the electrode, a potential well (depletion area) is formed in the silicon substrate directly beneath the gate. The absorption of incident photons generates electron-hole pairs and the free electrons generated in the vicinity of the capacitor are stored and integrated in the potential well. The number of electrons in the potential well is a measure of the incident light intensity which is detected at the edge of the CCD area by transferring the charge package according to the principle of charge transfer through the propagation of potential wells.

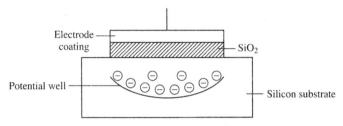

Figure 12.9 Structure of a CCD pixel.

12.6
PLASMA DISPLAY

Basic element of a plasma display is a tiny gas discharge cavity of dielectric layers with the transparent electrodes outside. The cavity coated with phosphor material has a width about 100 μm and is filled with gases like neon or xenon at a pressure about 500 mbar. The discharge is maintained by applying ac voltage about 90 V (starting voltage is about 150 V ac). When an electric current is passed through the cell, the gas atoms, ions and so on are excited by electron collision, which in turn relaxes to the ground state radiatively. This (UV) radiation will absorb by the phophor coating causing light to emit in the visible range. To obtain colour display three such units coated with blue, green and red emitting phosphors are used to form a pixel. The voltage applied across each sub pixel will cause its characteristic colour. A large plasma display is formed by sandwitching such a large number of cells as pixels in between electrodes and the glass plate. The chief advantages of the plasma display are:

1. The screen is large with a thin size.
2. Very high brightness.
3. Large viewing angle.

REVIEW QUESTIONS

1. Explain the working of a Cathode Ray Tube.
2. What are liquid crystals? Give some examples.
3. Explain different liquid crystal structures.
4. Explain the working of an LCD.
5. Write short notes on plasma display.

13

Lasers

13.1
INTRODUCTION

One of the most significant development in the field of science and technology in the last century is the advent of lasers. Since their invention in 1960, lasers have been used for numerous applications in science, engineering, medicine, defence, communications and so many using a wide variety of lasers. Laser is a source of electromagnetic radiation having certain special properties. The word laser is used for representing Light Amplification by Stimulated Emission of Radiation.

13.2
BASIC PRINCIPLE OF LASER ACTION

The energy state of an atomic system is characterized by discrete energy levels. Usually an atom will be in the ground state, which can be excited to a higher energy state by a process known as absorption (also called as stimulated absorption; Figure 13.1). On the other hand, an atom in an excited state come to a lower energy state by means of emitting electromagnetic radiation by two processes. An atom in the excited state emit spontaneously even in the absence of any incident radiation. This process is referred to as spontaneous emission and depends on the number of atoms in the excited state. An incident radiation of appropriate frequency can also stimulate an atom in an excited state to emit radiation. This process is known as stimulated emission. The rate of stimulated emission depends on the intensity of the external field and also the number of excited state atoms. The net stimulated transition is proportional to the difference in the number of atoms in the excited and lower states.

Let N_1 and N_2 represent the number of atoms per unit volume in the lower and upper energy states with energies E_1 and E_2 respectively (Figure 13.2). An electron can be excited by absorbing a photon of frequency

$$v = \frac{E_2 - E_1}{h}$$

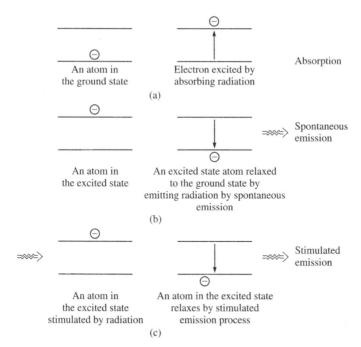

Figure 13.1 Principle of absorption spontaneous emission and stimulated emission.

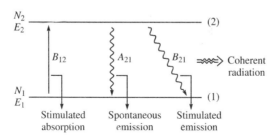

Figure 13.2 Laser action in a two level laser system.

This upward transition is called stimulated absorption or simply absorption. (Absorption rate coefficient is B_{12}). The transition from the upper state to the lower state can be by two ways spontaneous emission (A_{21}) and by stimulated emission (B_{21}).

At thermal equilibrium the total upward transition rate due to absorption is equal to the total downward transition rate due to both the spontaneous and stimulated emission.

The upward transition rate due to absorption = $N_1 B_{12} \rho(v)$

The spontaneous emission transition rate = $N_2 A_{21}$ and

The stimulated emission transition rate = $N_2 B_{21} \rho(v)$

∴ $$N_1 B_{12} \rho(v) = N_2 A_{21} + N_2 B_{21} \rho(v)$$

$$\therefore \quad \rho(v) = \frac{A_{21}/B_{21}}{[(N_1/N_2)(B_{12}/B_{21})] - 1}$$

where $\rho(v)$ is the radiation energy density given by the Planck's law of black body radiation

$$\rho(v) = \frac{8\pi v^2}{c^3} \frac{hv}{\exp(hv/kT) - 1}$$

According to Boltzmann statistics, the population of two levels at a temperature T K at thermal equilibrium are related by

$$\frac{N_1}{N_2} = \exp\left(\frac{E_2 - E_1}{kT}\right) = \exp\left(\frac{hv}{kT}\right)$$

Thus $\quad \dfrac{A_{21}}{B_{21}} = \dfrac{8\pi h v^3}{c^3} \quad$ and $\quad B_{12} = B_{21}$

where A and B are known as Einstein A, B coefficients.

At thermal equilibrium ratio of the rate of spontaneous emission to the rate of stimulated emission is related by

$$\frac{A_{21}}{\rho(v)B_{21}} = \exp\left(\frac{hv}{kT}\right) - 1$$

At thermal equilibrium for $v \ll kT/h$ the stimulated emission far exceeds the spontaneous emission, while for $v \gg kT/h$ the spontaneous emission far exceeds the stimulated emission. Thus if the rate of stimulated emission is much higher than the rate of spontaneous emission then the number of atoms in the excited state is higher than that in the lower state. To achieve laser action, it is necessary to have $N_2 > N_1$, and such a condition is known as population inversion.

When a light beam passes through an absorbing medium, the intensity as a function of distance is given by

$$I(z) = I_0 \exp(-\alpha z)$$

where α is the absorption coefficient of the medium given by

$$\alpha = \left(\frac{g_2}{g_1} N_1 - N_2\right) \frac{Bhv\mu}{c}$$

where g is the degeneracy of the state, μ is the refractive index of the medium. Hence, if $N_2 > N_1$ the absorption coefficient is negative known as small signal gain coefficient K, then

$$I(z) = I_0 \exp(Kz)$$

Thus, there is an exponential increase in intensity of the beam, as it propagates through the medium so that light gets amplified. This is the basic principle of laser action.

13.3
PARTS OF A LASER SYSTEM

There are three main components for a laser system (Figure 13.3). They are:
1. Active medium
2. Pumping mechanism
3. Optical resonator (or cavity)

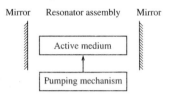

Figure 13.3 Components of a laser system.

The active medium consists of a collection of atoms, molecules or ions in the form of solid, liquid or gas, which act as a medium for amplification of light waves. The active medium must have the capability to produce population inversion under suitable condition. Depending on the active medium used, lasers can be broadly categorized into solid, liquid or gaseous lasers, where the active material is in the form of solid, liquid or gaseous form, respectively. The frequency of the laser is determined by the energy levels of the atoms, ions or molecules.

For light amplification, the medium has to be excited and has to be kept in a state of population inversion. The process by which the atoms are raised from a lower level to an upper state is called pumping process. As the active medium has a large number of energy levels with complex excitations paths, different mechanisms of excitations have been employed in lasers. Some of these excitation mechanisms are by strong light, say flash lamp—known as optical pumping, by electron impact—known as electrical pumping, by chemical reaction—known as chemical pumping, by means of supersonic gas expansion—known as gas dynamic pumping, etc. The selection of the type of pumping depends on the active material involved to achieve population inversion. Thus, the pumping mechanism is provided to obtain a state of population inversion between a pair of energy levels of the atomic system.

As the excited atoms decay spontaneously by emitting photons at random, to obtain a coherent radiation it is necessary to restrict the number of photon states. This can be achieved by enclosing the active medium in a resonator (cavity) tuned the frequency concerned. For example, two plane or curved mirrors—one of which has 100% reflection and the other which has partial reflection (>90%) is a suitable arrangement for an optical resonator. If the active medium is placed in between two such mirrors, the emitted light reflect back and forth in the resonator assembly, and hence, provide a feedback for light amplification.

The output of a laser can be either in pulsed or continuous wave mode. In continuous wave mode, the output will be continuous, whereas in pulsed mode output will remain only for a short duration at a certain repetition rate. The properties of the laser beam—line width, optical power, spectral characteristics,

13.4
PROPERTIES OF A LASER BEAM

Important properties of a laser beam are:
1. High monochromaticity
2. High intensity
3. High degree of coherence
4. High directionality

Monochromaticity
The light emitted from a laser has single colour or a single wavelength with a small spectral width, i.e. a laser beam is highly monochromatic.

Directionality
The spread (diverging) of a laser beam is very small as compared with the ordinary light, i.e. the light from a laser is highly directional. Hence, a laser beam can travel a long distance without much divergence.

High Intensity
Because of its high directionality a laser beam can be focussed to an extremely small spot size which will give rise to a very high intensity.

Coherence
A light beam is said to be coherent if all the waves are in same phase, same amplitude and same wavelength. It is a measure of the degree of ordering of the light field. A laser beam that has a high spatial and temporal coherence implies that a high degree of the ordering of light field with respect to space and time.

All lasers have these properties, and it is these properties that made them a versatile tool for various applications.

13.5
OPTICAL RESONATOR

In a laser system with an active medium which is capable of light amplification a part of the output energy must be coupled back into the system. Such a feedback is provided by placing the active medium in between a pair of mirrors which are facing each other. Such a system formed by a pair of mirrors is referred to as an optical resonator. Initial stimulus is provided by spontaneous transition between the appropriate energy levels and when the gain provided by the medium exactly matches the loses during a complete round trip, saturation may be reached.

Commonly used laser resonator configurations are given as follows:

13.5.1 Parallel Plane Resonator

In Figure 13.4, two plane mirrors are placed at a distance d apart with reflecting surfaces facing each other. The mirrors are flat within $\lambda/100$ and should be parallel within about one second of an arc.

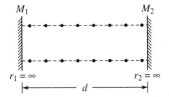

Figure 13.4 Plane parallel resonator.

13.5.2 Confocal Resonator

Confocal resonator system consists of a pair of spherical mirrors of same radii of curvature $r_1 = r_2 = r$ separated by a distance equal to the radius of curvature (see Figure 13.5).

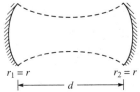

Figure 13.5 Confocal resonator.

13.5.3 Spherical Resonator

A general resonator system consists of a pair of spherical mirrors of radii of curvature r_1 and r_2 ($r_1 \neq r_2$) which are separated by a distance d as shown in Figure 13.6.

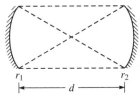

Figure 13.6 Spherical resonator.

13.5.4 Large Radius Resonator

A large radius resonator system has two mirrors with radii of curvature r_1 and r_2, and they are arranged such that $r_1, r_2 \gg d$ (Figure 13.7).

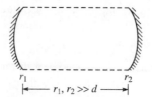

Figure 13.7 Large radius resonator.

13.5.5 Hemispherical Resonator

A hemispherical resonator has a plane mirror and a spherical mirror of radius of curvature d which are separated by a distance d (Figure 13.8).

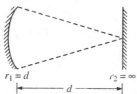

Figure 13.8 Hemispherical resonator.

In Figures 13.4 and 13.8 dotted lines show the extend of the mode volume in each case. In a stable resonator, a ray of light may keep bouncing back and forth between the mirrors without escaping from the system. On the other hand, in an unstable resonator system, the rays diverge away from the axis and escape from the resonator after a few traversals. The gain is usually very small so it is very essential to minimize the losses that occurred in the system. In order to sustain oscillations in the cavity it requires net losses suffered by the beam must be compensated by the gain of the medium.

13.6
MODES OF A PLANAR OPTICAL RESONATOR

In an optical resonator, field configuration is represented by standing wave pattern which is known as mode of oscillation. In comparison to the wavelength of light, optical resonators have a large dimension so that in general these will be a large number of modes can oscillate within the laser cavity. To obtain a highly directional spectrally pure output, various techniques have been employed to select a particular mode to oscillate with high gain.

In a planar optical resonator, with two plane mirrors separated by a distance L apart, standing wave patterns are set up when

$$m = \frac{L}{\lambda/2}$$

where m is the mode number and λ is the wavelength of the longitudinal modes. The frequency of oscillation is given by

$$v_m = \frac{mV}{2L}$$

where V is the velocity of light within the active medium. The longitudinal mode spacing which is equal to the difference in frequency or spectral width is

$$v_{m+1} - v_m = \frac{V}{2L}$$

or

$$\Delta v = \frac{V}{2L}$$

13.7
QUALITY FACTOR OF AN OPTICAL RESONATOR

Total losses in a resonator system are due to a number of different processes as follows:
1. Absorption and scattering of radiation at the cavity mirrors.
2. Absorption in the medium due to transitions other than the desired transitions
3. Scattering due to inhomogeneities and impurities present in the medium
4. Diffraction losses at the mirrors
5. Transmission through the cavity mirrors.

The dissipation of energy in an optical resonator is described in terms of quality factor Q of the mode. The Q-factor is defined as

$$Q = \frac{2\pi v_0 \times (\text{energy stored in the mode})}{\text{Energy dissipated per second in that mode}}$$

$$= \frac{2\pi v_0 W(t)}{-dW(t)/dt}$$

$$\frac{dW(t)}{dt} = -\frac{2\pi v_0}{Q} W(t)$$

Integrating

$$\int_{W(0)}^{W(t)} \frac{dW}{W} = -\frac{2\pi v_0}{Q} \int_0^t dt$$

or

$$W(t) = W(0) \exp\left(\frac{-2\pi v_0}{Q} t\right)$$

v_0 represents the oscillation frequency, $W(t)$ is energy stored in the mode at time t. The energy stored in a mode decays exponentially with time, and decays to $1/e$ times the value of W at $t = 0$, i.e.
When $t = t_c$, $W(t_c) = W(0)/2$

where

$$t_c = \frac{Q}{2\pi v_0}$$

is called the cavity lifetime. For a Lorentizian line shape function the full width at half maximum is inversely proportional to the quality factor

$$\Delta v = \frac{v_0}{Q}$$

The time taken by light for one complete traversal within the cavity is

$$t = \frac{\text{Distance}}{\text{Velocity}} = \frac{2L}{V} = \frac{2L\mu}{c}$$

where μ is the refractive index of the medium. Thus

$$W(t) = W(0)\exp\left(\frac{-2\pi v_0}{Q} \cdot \frac{2L\mu}{c}\right)$$

If γ is the absorption coefficient per length and R_1 and R_2 are the reflection coefficient, the energy remaining within the cavity after a complete traversal is

$$W(t) = W(0)\, R_1 R_2 \exp(-2\gamma L)$$

Thus

$$W(0)\, R_1 R_2 \exp(-2\gamma L) = W(0)\exp\left(\frac{-4\pi v_0}{Q} \cdot \frac{\mu L}{c}\right)$$

or

$$2\gamma L = \log(R_1 R_2) + \frac{4\pi v_0 \mu L}{cQ}$$

or

$$Q = \frac{4\pi v_0 \mu L}{c} \cdot \frac{1}{2\gamma L - \log(R_1 R_2)}$$

13.8
THREE LEVEL LASER SYSTEM

When an active medium in a laser system is excited by pumping mechanism, atoms get excited and the population in various levels may change. In a three level system, various excitation and de-excitation process can be expressed by a simplified model involving three energy levels E_1, E_2 and E_3. The level 1 denotes the ground state and the levels 2 and 3 denote two excited states as shown in Figure 13.9. The pumping transfers population from level 1 to level 3, which then decays to level 2 by fast non-radiative processes. Laser emission corresponds to a transition between energy levels from E_2 to E_1. Populations in the levels 1, 2 and 3 are denoted by N_1, N_2 and N_3, respectively.

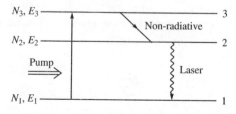

Figure 13.9 Energy levels in a three level laser system.

To achieve population inversion between the levels 2 and 1, $N_2 - N_1$ must be positive. This occurs when the level 2 is a metastable state with a long lifetime, i.e. the lifetime of the level 2 is larger than that of the level 3.

13.9
FOUR LEVEL LASER SYSTEM

Energy level diagram of a four level laser system is shown in the Figure 13.10. Level 1 represents the ground state, the levels 2, 3 and 4 represent the excited states with population density N_1, N_2, N_3 and N_4, respectively in each state. In general, the level 4 is a broadband (i.e. a collection of energy levels) and the pumping transfers a large population to 4 from the level 1, the population then decays to the lower level 3 by non-radiative processes. The population inversion is established between levels 3 and 2, and the laser actions take place at a frequency corresponding to $v = (E_3 - E_2)/h$. The population of 2 is then non-radiatively decay to the ground state. The lower laser level 2 is required to be sufficiently above the ground state, so that at ordinary temperatures, the population of level 2 is negligible. To achieve population inversion in a four level laser system $N_3 - N_2$ must be positive. For that it requires the lifetime of the level 3 which must be larger than that of the level 2. A four level system is more efficient than a three level laser system.

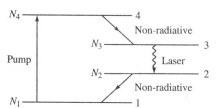

Figure 13.10 Energy levels in a four level laser system.

13.10
RUBY LASER

A ruby laser consists of a rod of ruby (~5 cm in length and 0.5 cm in diameter). It is crystal of aluminium oxide, with about 0.05 per cent by weight of aluminium atoms replaced by chromium atoms. The Cr^{3+} present as impurities in Al_2O_3 crystal is the lasing material. The two ends of the ruby rod are made plane parallel and polished to form as mirrors (M_1 and M_2). One of the end is fully reflecting and the other is partially reflecting (~10% transmission), so that they act as an optical resonator. In a typical set up ruby is surrounded by a helical flash lamp as shown in Figure 13.11. The flash lamp is connected to a capacitor which is charged by applying a high voltage. The optical pumping is performed by subjecting the ruby rod to intense light flashes from the xenon lamp.

184 Photonics: An Introduction

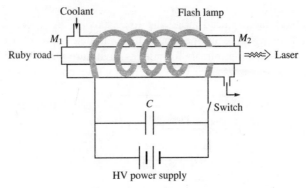

Figure 13.11 Ruby laser.

Energy levels of the chromium ions in the ruby crystal that generates laser is shown in Figure 13.12. When the high voltage is applied by switching on, the capacitor will be charged and when the switch is off, the capacitor discharges so that the lamp flashes with an intense pulse with a few thousands of Joules of energy in a duration of 5×10^{-4} seconds. A part of the energy is absorbed by the ground state Cr^{3+} ions, thereby transferred to the bands denoted by E_1 and E_2 (absorbing at 400 nm and 600 nm). These absorption bands have small lifetime of the order of 5×10^{-8} s. From these bands they suffer a very fast non-radiative decay to the metastable state M having a long lifetime of 3×10^{-3} s. The number of atoms in this state goes on increasing and leads to population inversion between this metastable state and the ground state. Ruby laser is a typical example of a three level laser system. The laser wavelength corresponds to $\lambda = 694.3$ nm and at room temperature the linewidth of the transition is about 11 cm^{-1}.

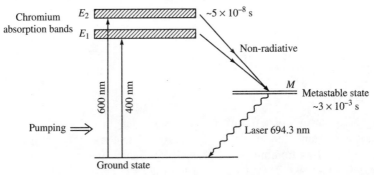

Figure 13.12 Energy levels of Cr^{3+} in ruby laser system.

The first laser system developed was the ruby laser, which was by Maiman in 1960.

13.11
Nd:YAG LASER

Nd:YAG laser is one of the most important laser system in which trivalent neodymium (Nd^{3+}) ions are present as impurities in yitrium aluminium garnet

(YAG; $Y_3Al_5O_{12}$) lattice. A typical arrangement consists of a Nd: YAG laser rod and a flash lamp placed along the focal axis of an elliptical reflector. This configuration helps to focus maximum light emitted from the flash lamp to the laser rod. The reflecting mirrors forming as cavity are placed outside the elliptical reflector. Nd: YAG lasers are usually excited by krypton flash lamp (Figure 13.13).

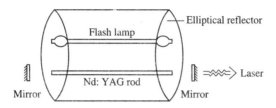

Figure 13.13 Nd: YAG laser system.

The relevant energy levels are shown in Figure 13.14. Optical pumping excite the Nd^{3+} ion and create a large population in a number of bands between 13000 to 25000 cm^{-1}, which then de-excite by non-radiative processes to the upper laser level $^4F_{3/2}$. The lifetime of this level is about 0.23×10^{-3} s (a metastable). The laser transition corresponds to $^4F_{3/2} \rightarrow {}^4I_{11/2}$ at $\lambda = 1.06$ µm. Transition from $^4I_{11/2}$ to the lowest state take place by fast non-radiative processes (Figure 13.14). This system is a typical example for a four level system which can produce an intensity as high as 10^{16} watts/cm^2 with pulse width about picoseconds.

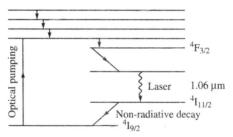

Figure 13.14 Laser line in Nd^{3+}: YAG laser.

13.12

He–Ne LASER

A typical helium–neon gas laser consists of a quartz gas discharge tube about 80 cm in length and 1 cm in diameter (see Figure 13.15). Ends of the discharge tube are made with optical windows kept at brewsters angle. The tube is filled with a mixture of helium and neon (about 1 mm of Hg of helium gas and 0.1 mm of Hg of neon gas) which acts as the lasing medium. Two electrodes are provided in the envelope and the discharge is excited by applying a dc high voltage of power about a few tens of watts. The tube is placed in between a pair of mirrors forming an external cavity, in which one of the mirror is completely reflecting and the other partially reflecting to couple out the laser beam.

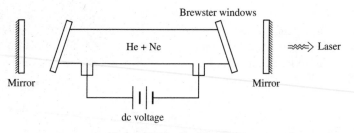

Figure 13.15 He–Ne laser.

When a discharge of He and Ne is established by applying a dc voltage, the accelerated electrons collide with helium atoms and excite them to the higher energy states. The He atoms are easily excited by electron impact than the Ne atoms. The simplified energy level diagram of He and Ne is shown in Figure 13.16.

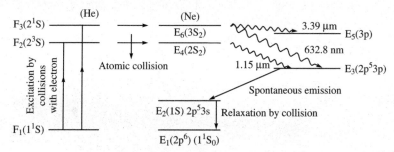

Figure 13.16 Energy level diagram of He and Ne.

Many of the helium atoms are collected in the long lived metastable states $F_2(2^3S)$ and $F_3(2^1S)$ whose lifetimes are 10^{-4} s and 5×10^{-6} s, respectively. Since these long lived levels nearly coincide in energy with $E_4(2S)$ and $E_6(3S)$ levels of neon, they can excite ground state neon atoms into these two excited states. This excitation takes place when an excited helium atom collides with a neon atom in the ground state and exchange energy between them via following reactions.

$$He(2^3S_1) + Ne(1^1S_0) \rightarrow He(1^1S_0) + Ne(2S_2)$$
$$He(2^1S_0) + Ne(1^1S_0) \rightarrow He(1^1S_0) + Ne(3S_2)$$

This pumping mechanism in the He–Ne system provides a selective population of 2S and 3S levels of Ne atoms.

Because of the long lifetimes of the levels F_2 and F_3 of helium, the process of energy transfer has a high probability. Thus, the discharge through the gas mixture continuously populates the neon excited levels E_4 and E_6. This helps to create a state of population inversion between the levels E_4 (or E_6) and the lower lying energy levels E_5 and E_3. The lifetime of S level is of the order of 100 ns while that of p-levels is of the order of 20 ns. The population inversion can be achieved leading to laser emission at wavelength 3.39 μm ($E_6 \rightarrow E_5$), 1.15 μm ($E_4 \rightarrow E_3$) and 632.8 nm ($E_6 \rightarrow E_3$). He–Ne laser is a typical example of a four-level laser system. Light from He–Ne laser is more monochromatic and directional as compared to solid state lasers. (This is due to various crystalline imperfections present in the solid, scattering and heating due to the flash lamp). The 632.8 nm He–Ne laser is one of the most popular and almost widely used laser. Presently, He–Ne lasers with compact size of length a few centimetres having an output power of 0.5 mW to 50 mW are available.

13.13
CO_2 LASER

CO_2 laser is a molecular laser, in which a mixture of CO_2, N_2 and He or water vapour are used. The active centres are the CO_2 molecules. This laser is very efficient compared with other gas lasers, and gives a high power from several watts to several kilowatts of optical power. They are inexpensive and easy to construct.

When a discharge is produced in a tube containing above gases, the molecules are excited to the higher electronic, vibrational and rotational states by electron impact. Relevant energy levels of CO_2 and N_2 leading to population inversion is shown in Figure 13.17. A large fraction of the CO_2 molecules excited by electron impact cascade down and are collected in a long lived upper laser level denoted by (001). N_2 molecules that are excited by electron impact are collected in the upper state of N_2, and then they collide with ground state CO_2 molecules, resulting to excite CO_2 ground state atom to the upper laser state by energy transfer. This process maintains sufficient population in the upper level. The laser emission corresponds to transition between the rotational sublevels of upper vibrational band and rotational sublevels of a lower vibrational band of the CO_2 molecule. The population inversion between the upper level (001 states) and lower levels (021 and 100 states) leads to oscillations at 9.6 and 10.6 μm. At partial pressures CO_2 = 15. Torr, N_2 = 1.5 Torr and H_2 = 12 Torr, the life time of the upper level is approximately 0.4 ms and that of the lower levels are 20 μs. N_2 and He gases increase the efficiency of the laser, in which N_2 helps to increase the population of the upper level and to depopulate the lower level keeping CO_2 cold by conducting heat away to the walls.

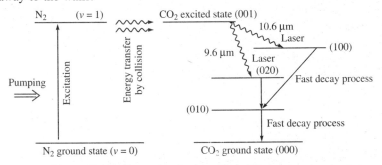

Figure 13.17 Energy level scheme in a CO_2 laser system.

Arrangement of a typical CO_2 laser is shown in Figure 13.18. It consists of a discharge tube of glass of length about 1 m and 2.5 cm in diameter. Ends of the tube are made with alkali halide brewster windows. Two confocal silicon mirrors coated with aluminium are placed externally, act as resonant cavity. A gas mixture of CO_2, N_2 and He at an optimum partial pressure is maintained in the tube. To remove dissociations products, a continuous flow of the gas mixture is maintained. The gas is excited by applying a high dc voltage. The laser action produces output at wavelength 10.6 μm (001 → 100) and at 9.6 μm (001 → 020).

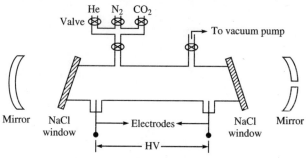

Figure 13.18 CO_2 laser system.

13.14

DYE LASER

The active medium in a dye laser is a liquid medium. In dye lasers, an organic dye, which is an organic compound that absorb strongly in certain wavelength (visible, IR) region is dissolved in a suitable liquid such as ethyl alcohol, methyl alcohol, acetone, benzene, toluene, water and so on are used as active medium. Lasing range 0.3 μm to 1.3 μm can be covered by using many organic dyes which are classified into xanthenes, polymethines, oxaines, coumarins, anthracenes, azines, acridines and phthaloziamin group. The most widely used organic dye is Rhodamine 6G (xanthene group) which emits in yellow-red region.

A schematic representation of energy levels of an organic dye molecule is shown in Figure 13.19. The state S_0 is the ground state, S_1, S_2, T_1 and T_2 are the excited electronic states of the molecules. Typical energy separation between them is about 20,000 cm^{-1}. The singlet(S) and triplet(T) electronic states are further split into vibrational and rotational levels with energy separation about ~1500 cm^{-1} and ~15 cm^{-1}, respectively. The transition between singlet-singlet or triplet-triplet are strongly allowed, whereas singlet-triplet or triplet-singlet transitions are weakly allowed.

In the process of pumping, the molecules are first excited to a vibrational state of S_1. Most of the dye molecules non-radiatively drop down to the lowest energy level of S_1. Fluorescence is emitted when molecule decay from these to any vibrational sublevel of S_0.

Molecules from the states S_1 can also make a non-radiative relaxation to the triplet level T_1, by a process referred to an intersystem crossing. The intersystem crossing will reduce the gain since it will deplete the number of molecules available in the upper state for laser action due to following reasons:

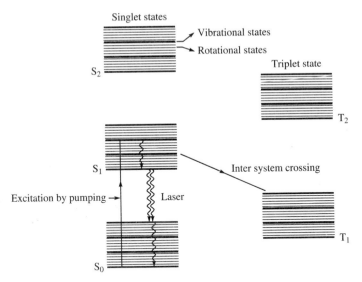

Figure 13.19 Energy levels of a dye molecule.

1. Causes to the reduction in the population in S_1 level due to $S_1 \to T_1$ transition.
2. Since the transition $T_1 \to T_2$ is allowed and the wavelength in this absorption is almost equal to that corresponding to the fluorescence transition $S_1 \to S_2$.

The experimental set up of a dye laser with wavelength tunability is shown in Figure 13.20. The dye solution is taken in a quartz cuvette which is then optically excited by a flash lamp. A mirror and a reflection grating are used to form optical cavity. Wavelength selection can be accomplished by rotating the grating.

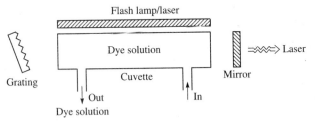

Figure 13.20 Dye laser system.

13.15
SEMICONDUCTOR LASER

In a semiconductor laser system, laser beam is obtained by creating population inversion in a semiconductor *pn* junction. When a *pn* junction is forward biased, holes are injected from *p*-region to the *n*-region and the electrons from *n*-region to the *p*-region. They combine with carriers of opposite charges in the depletion region and release excess energy in the form of heat or light. In semiconductor lasers, most of the excess energy is released in the form of light through the junction. The current density required is sufficiently high.

Semiconductor lasers are extremely small in dimension. Resonant cavity is formed by cleaving the end surfaces and cutting the crystal such that two ends are perpendicular to the junction, and parallel to one another. The population inversion that is necessary for amplification of radiation is in a narrow region near the junction called the depletion region (Figure 13.21).

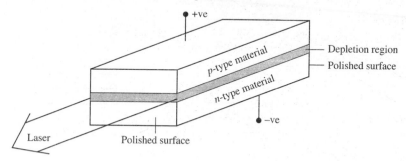

Figure 13.21 Semiconductor diode laser.

Homojunction semiconductor lasers usually require a very low temperature below liquid nitrogen, whereas hetrojunction semiconductor lasers can be operated continuously at room temperatures. The wavelength of the light emitted, depends on the bandgap of the semiconductor material which falls from UV to IR range. GaAs, GaInAsP, etc, and are some of the commonly used semiconductor laser materials.

13.16
APPLICATIONS OF LASERS

Lasers have tremendous applications in the field of science, technology, medicine, defence etc., by utilising the properties such as high power, coherence, monochromaticity, directionality, tunability, etc. Some important applications are listed as follows:

1. Due to their high power and the fact that the laser beam can be focussed to an extremely small spot size they find wide applications in industry for cutting, welding, drilling, material processing, positioning, monitoring etc. In electronic industry, they are used for etching, soldering, material processing and so on.

2. In science, they are used for spectroscopy, atomic, molecular structure studies, isotope analysis and separation, laser produced plasma, chemical analysis, etc.

3. They also find applications for non-linear optical processes like self-focussing, Raman effect, second harmonic generation, parametric oscillations, optical phase conjugation, ultra fast phenomena, and so on.

4. Laser can be used to initiate fusion reaction by focussing high power laser beams.
5. They are used for holography—a method for producing three-dimensional images—image processing, data storage, information processing, printing etc.
6. Optical communication is an important field of application using lasers.
7. In defence, it is used to guide missile, rockets, etc.
8. Light detection and ranging accurate measurement of distances and sizes, tracking, etc.
9. In medical field, it is used for painless surgery, cancer treatment eye surgery and so on.

EXAMPLE 13.1 In a He–Ne laser, the laser line corresponds to a transition between the energy states 166659 cm^{-1} and 150856 cm^{-1}. Calculate the wavelength of the laser. If the power is 5 mW, calculate the number of photons emitted per second.

Solution $E_1 = 150856$ cm^{-1}, $E_2 = 166659$ cm^{-1}, $P = 5 \times 10^{-3}$ W, $\lambda = ?$, $n/t = ?$.

$$E_2 - E_1 = 15803 \text{ cm}^{-1}$$

∴
$$\lambda = \frac{1}{15803} = 6328 \text{ Å}$$

$$P = \frac{n}{t} h\nu$$

∴
$$\frac{n}{t} = \frac{P\lambda}{hc} = \frac{5 \times 10^{-3} \times 632.8 \times 10^{-9}}{6.626 \times 10^{-34} \times 3 \times 10^8}$$

$$= 1.59 \times 10^{16} \text{ photons/s}$$

EXAMPLE 13.2 Half width of a spectral line at 632.8 nm is 2×10^{-3} nm. Calculate the half width in frequency.

Solution
$$c = \nu\lambda$$

$$\nu = \frac{c}{\lambda}$$

∴
$$d\nu = \frac{-c}{\lambda^2} d\lambda$$

∴
$$|d\nu| = \frac{3 \times 10^8}{(632.8 \times 10^{-9})^2} \times 2 \times 10^{-3} \times 10^{-9}$$

$$= 1.5 \text{ GHz}$$

EXAMPLE 13.3 For a planar resonator, calculate the length of the resonator if the longitudinal spacing is 150 MHz.

Solution
$$\Delta\nu = \frac{V}{2L}$$

$$\therefore \quad L = \frac{V}{2\Delta v}$$

$$= \frac{3 \times 10^8}{2 \times 150 \times 10^6} = 1 \text{ m}$$

EXAMPLE 13.4 For He–Ne laser with wavelength 632.8 nm has a line width 1 MHz. Calculate the Q factor.

Solution

$$\Delta v = \frac{v_0}{Q}$$

$$\therefore \quad Q = \frac{c/\lambda}{\Delta v}$$

$$= \frac{3 \times 10^8}{1 \times 10^6 \times 632.8 \times 10^{-9}} = 4.7 \times 10^8$$

REVIEW QUESTIONS

1. Explain with energy level diagram the absorption, spontaneous emission and stimulated emission processes.
2. Explain the basic principle of laser action.
3. What do you mean by population inversion?
4. What are A and B coefficients?
5. Explain the main components of a laser system.
6. Explain important properties of a laser beam.
7. Explain different configuration of optical resonators.
8. Write a short note on the modes of a planar waveguide.
9. Derive the expression for Q factor of an optical resonator.
10. What do you mean by Q factor?
11. What are the different types of losses in an optical resonator?
12. Write short notes on three level and four level laser system.
13. Explain the working of the following laser systems:
    (i) Ruby laser
    (ii) Nd:YAG laser
    (iii) CO_2 laser
    (iv) He–Ne laser
    (v) Dye laser
    (vi) Semiconductor laser
14. Give some important applications of lasers.

14

Advances in Photonics

This chapter briefly discusses some of the advanced topics related to photonics.

14.1
RAMAN SCATTERING

Raman effect is an inelastic process leading to scattering of light by molecules of solid, liquid or gaseous medium in which the scattered radiation consists of lower and higher frequency components in addition to the incident frequency. This phenomenon was discovered by C.V. Raman in 1928. The medium which exhibits this property is known as Raman active medium. The scattered light components with lower frequency are known as Stokes lines and those with higher frequency components are known as anti-Stokes lines.

Energy level diagram describing two types of Raman processes with respect to the vibrational levels of molecules are shown in Figure 14.1. The incident radiation excite the ground state molecule (v = 0), which then emit a Stoke photon

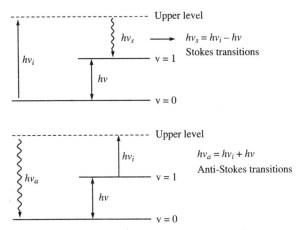

Figure 14.1 Energy level diagram describing Stokes and anti-Stokes lines in Raman effect.

of frequency $v_s = v_i - v$, by radiative relaxation to the vibrational state v = 1 (which is an excited state of the molecule). The frequency of the Stoke photon is less than that of the incident photon, such that

$$hv_s = hv_i - hv$$

In the second process, v = 1 level of the molecule is excited by the incident photon at v_i and de-excite to the ground state v = 0, emitting an anti-Stokes photon of frequency v_a such that

$$hv_a = hv_i + hv$$

The frequency of the anti-Stokes photon is greater than that of the incident photon.

The energy lost or gained ($\mp hv$) by the electromagnetic field is a characteristic property of the molecules in the Raman active medium so that it is a versatile analytical tool to detect the composition or the structural properties of the molecules. The anti-Stokes lines are very weak as compared to the intensity of the Stokes lines since the population of v = 1 is less than that of v = 0 state at thermal equilibrium.

To observe the Raman effect, the Raman active medium is illuminated by an intense light source such as a mercury lamp or laser and the scattered light is observed spectroscopically as in Figure 14.2. Analysis reveals the presence of spectral lines with frequencies shifted to lower and higher side along with the unscattered incident line.

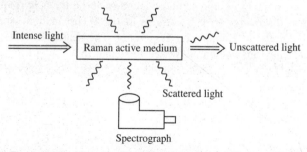

Figure 14.2 Experimental aspects to observe Raman effect.

Using an intense laser beam as an exciting source a stimulated version of the Raman scattering can be achieved which is more efficient than the spontaneous Raman scattering, since more incident photons are converted into Stoke and anti-Stoke photons. In this case, the photon scattered by spontaneous processes acts as a seed photon, which inturn stimulates Raman process. Stimulated Raman scattering leads to an intense cone-shaped beam both in the forward and backward directions while spontaneous Raman process produces a weak isotropic emission.

14.2
PHOTOREFRACTIVE EFFECT

Photorefractive effect is the change in index of refraction of certain crystals due to the redistribution of charges (electrons or holes) caused by optical irradiation.

When an optical material exhibiting photorefractive effect is illuminated by two light beams of same frequency, they interfere within the crystal volume so that the light intensity is modulated as shown in Figure 14.3. Charges are produced within the region by photoionization, the density of which is proportional to the modulated light intensity. These carriers are then diffuse or drift within the crystal and redistribute themselves and a periodic space charge is set up. This spatial charge distribution will create a spatially varying electric field distribution, which inturn will modulate the refractive index via linear electro optic effect as shown in the figure.

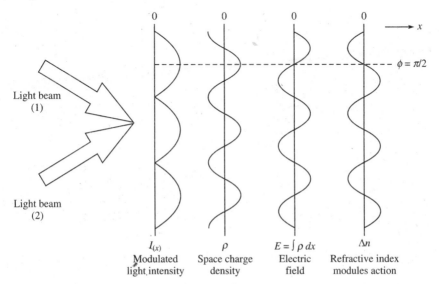

Figure 14.3 Principle of photorefractive effect.

The change in index of refraction produced is given as

$$\Delta n = -\frac{1}{2} n_0^3 r_{\text{eff}} E$$

where n_0 is the refractive index of the crystal, r_{eff} is the electro optic coefficient that depends on the crystal orientation and the field direction, E is the low frequency electric field. Note that there is a phase shift of $\pi/2$ to the modulated refractive index with respect to the modulated light intensity and this phase shift leads to a transfer of energy between the two incident beams.

For example, if a photorefractive crystal is irradiated with a signal and pump beam, they will interfere to form a non-uniform intensity distribution and the non-linear response will develop a refractive index modulation. As a result of phase shift introduced between refractive index change and the light intensity, there will be an intensity redistribution in the output beam, the signal beam is amplified, whereas the pump beam is attenuated (Figure 14.4).

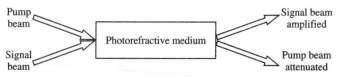

Figure 14.4 Photorefractive effect by two beam coupling.

The photorefractive effect gives rise to a strong optical non-linearity, and since diffusion and drift of charges are involved, the response of the effect is very slow of the order of 100 ms. In steady state, it is independent of light intensity that induces the photorefractive index change.

14.3

OPTOGALVANIC EFFECT

Passage of electricity through gases usually produces a great variety of luminous phenomena (glow, positive column etc. discharges) which is a characteristic of an electrical gas discharge. Electrical characteristics of the gas discharge can also modify by illuminating it with electromagnetic radiation. Optogalvanic Effect (OGE) is such a phenomenon in which impedance of a gas discharge is changed due to modification in population of states caused as a result of resonant absorption of electromagnetic radiation by atoms, ions, molecules or electrode material present in the discharge.

Absorption of photons by atoms, molecules, ions, etc., will significantly affect the relative population of various states in the discharge. This results a change in electron generation rate, which inturn causes a change in impedance of the discharge requiring a different voltage to sustain it. The optogalvanic signal can corresponds to an increase or decrease in the discharge current depending on the kinetics of the levels whose populations are perturbed by light. If the light excites atoms from a lower level with a lesser probability of ionization to a level with larger probability of ionization, the discharge current will increase. Alternatively, if the light excites atoms from a level with a larger probability of ionization to a level with a smaller probability of ionization, the discharge current will increase. This situation often arises when the lower level has a high probability of collisional ionization and the upper level is a short lived resonance level.

When the change in voltage (ΔV) across the cell called as OG signal, is proportional to the change in the local population of the excited states ΔN_i, then

$$\Delta V = \sum_i \alpha_i \Delta N_i$$

where $\alpha = \partial V / \partial N_i$; $i = 1, 2, \ldots$, etc.

This relation is based on the assumption that the variation in population for each state within the discharge volume depends on the relative optical absorption and the discharge temperature.

Several types of electrical gas discharges such as positive column, glow, hollow cathode, plane electrode, etc. discharges filled with an appropriate gas at low pressure can be used as the discharge medium. Discharge is maintained by applying a high voltage (dc) between two metal electrodes through a current limiting resistance. Under the influence of electric field, gas will ionize and most of the discharge properties are characterized by various excitation and de-excitation processes of the species present in the discharge. The modulated laser beam passes through the cell, the resonant absorption of radiation results into a change in impedance of the discharge, which can be measured using a cathode ray oscilloscope by blocking the dc voltage with a capacitor as shown in Figure 14.5. Pulsed or continuous wave laser can be used for exciting the discharge medium.

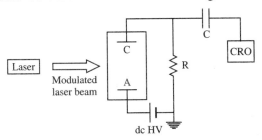

Figure 14.5 Set up for dc OGE effect.

The change in impedance of a gas discharge can also be generated by injecting electrons into the discharge via. photoelectric effect which is known as Photoelectron Optogalvanic Effect (POGE). In this non-resonant process, photon energy required is greater than the work function of the cathode material.

The photoelectrons emitted from the cathode will further interact with the discharge medium and the current in the circuit will be the resultant of the original plasma current and that produced by the interaction of photoelectrons within the plasma. The essential difference between OGE and POGE is that the former is due to the resonant absorption of radiation, whereas the latter is by the non-resonant absorption processes. OG effect has wide application in atomic and molecular spectroscopy, plasma diagnosis, high resolution spectroscopy, isotope studies, material characterization, etc.

14.4

PHOTOTHERMAL DEFLECTION EFFECT

A high intensity laser beam when falls on a material due to thermal effects refractive index will be modified. Now, if a probe beam is passed through it, the path of the emergent beam will be deflected (Figure 14.6). Such a deflection of the light beam due to modification in the refractive index caused by thermal effect is termed as photothermal deflection.

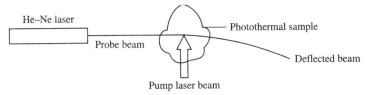

Figure 14.6 Photothermal deflection.

14.5

PHOTOREFRACTION IN A DIFFUSING MEDIUM

A beam of light when passes through a medium with a varying refractive index its path will deflect. In liquids such a deflection can be observed if there is a refractive index gradient in it. The refractive index gradient can be produced by mixing two soluble liquids of different refractive indices or by adding two or more liquids of different concentrations. The formation of refractive index gradient is mainly by diffusion of one liquid into the other. If a laser beam is passed through the refractive index gradient region, the deflection suffered to the beam is related to the diffusion.

The experimental set up to observe the photorefraction by diffusion is shown in Figure 14.7. The experimental cell is arranged at a suitable height. A fan of ray obtained by passing a laser beam (He–Ne) through a cylindrical lens is allowed to fall, diagonally on the cell along a straight line. Take 20 ml of water in the cell and pipetted 20 ml of the experimental solution (say NaCl) of known concentration into the bottom with no random mixing so that there is clear boundary separating two liquids. The laser beam passing through the mixing zone will deflect, the deflection being proportional to the refractive index gradient. As the variation of concentration within the medium in a plane is gaussian, the refractive index gradient, and hence deflection also follow the same pattern.

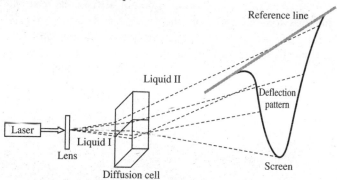

Figure 14.7 Photoreflective deflection in liquids by diffusion.

The deflected pattern is as shown in Figure 14.8, which can be traced out at a regular interval of time till the liquid mixes completely. This method is very suitable for studying diffusion mechanism of liquids.

Figure 14.8 Photorefractive deflection pattern.

14.6
SQUEEZED STATE

Consider an optical radiation with electric field represented by

$$E(t) = e \cos(\beta + \omega t)$$

where β is the phase and $e = E_1(t) + iE_2(t)$ is the complex amplitude of the electric field. Quantum mechanically, electric field cannot be specified exactly, so we can put it as

$$E_1(t) = E_{10} + \delta E_1(t)$$

and

$$E_2(t) = E_{20} + \delta E_2(t)$$

The corresponding uncertainties are

$$\Delta E_1(t) = \langle (E_1 - E_{10})^2 \rangle^{1/2}$$
$$= \langle [\delta E_1(t)]^2 \rangle^{1/2}$$

and

$$\Delta E_2(t) = \langle (E_2 - E_{20})^2 \rangle^{1/2}$$
$$= \langle [\delta E_2(t)]^2 \rangle^{1/2}$$

where $\langle \rangle$ implies the mean values. The uncertainty in the measurement of two components is

$$(\Delta E_1)(\Delta E_2) \geq \frac{A^2}{4}$$

where $A = (2\hbar\omega/\varepsilon V)^{1/2}$ is the absolute magnitude of the mean electric field amplitude for a single photon with energy $\hbar\omega$ inside a space of volume V with dielectric constant of the medium ε.

If $\Delta E_1 = \Delta E_2 = (A/2)$ uncertainty is equally divided between two quadrature components and said to be as coherent state. On the other hand, if $\Delta E_1 \neq \Delta E_2$, i.e. the uncertainties of the two quadrature components of the electromagnetic field are not equal, the field is said to be as squeezed state. Here the advantage is that it is possible to reduce the fluctuation of one of the component say ΔE_1 at the expense of the other ΔE_2 or vice versa keeping the product $(\Delta E_1)(\Delta E_2)$ at its initial value $A^2/4$. This is known as optical squeezing and can be achieved by non-linear phenomena such as degenerate optical parametric amplification. Some of the applications are photon counting, communication, atomic measurements, noise reduction in optical measurements, and so on. A pictorial representation of the coherent state and squeezed state is shown in Figure 14.9.

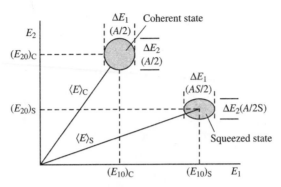

Figure 14.9 Pictorial representation of a coherent and a squeezed state. (S is the squeezing factor).

14.7
OPTICAL SOLITONS

When a light pulse propagates through a non-linear medium pulse shape can be modified as it propagates due to non-linear optical processes of Self Phase Modulation (SPM) and also by means of propagation effect such as Group Velocity Dispersion (GVD). SPM is the change in phase of an optical pulse resulting from non-linearity of refractive index of the medium. For a non-linear medium with intensity dependent refractive index $n = n_0 + n_2 I$ (n_0 is the intensity independent RI and n_2 is the non-linear coefficient, with n_2 positive), when an optical pulse travels the higher intensity portion of the pulse encounters a higher refractive index compared with the lower intensity region. Thus, the leading edge of the pulse exhibits a decrease in frequency, whereas the trailing edge exhibits an increase in the local frequency (i.e. we can observe a chirp in the pulse as it propagate). GVD leads to a broadening in the time domain, higher frequencies travel faster than the lower frequencies. If the medium exhibit GVD then it is possible that the frequency chirped pulse may be compressed under appropriate condition. The SPM is important for pulses with high peak intensity, whereas GVD is important for very short pulses. In general, these two effects may take place simultaneously and these two effects can cancel each other under certain circumstances so that the pulse can propagate a large distance with no change in shape. Such a self sustaining pulse propagating with no broadening at all in the time or frequency domain is known as optical soliton. As the solitons can propagate without broadening for a very large distance, they find very important application in optical fibre communication.

14.8
OPTICAL BISTABILITY

Bistability is a phenomenon in which the output can exists one of the two distinct stable state (low or high) irrespective of the input state. For small values of the input $I < I_1$ output is in a stable low state and for high input $I > I_2$ the output switches to the stable high state. As the input decreases, the output is also switched to the lower state via hysteresis as shown in Figure 14.10. The intermediate state is highly unstable a slight change in the input will trigger to output to the stable state.

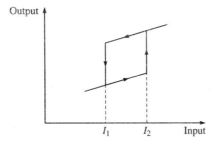

Figure 14.10 Bistability; input-output relationship.

Optical bistable devices exhibit such a bistable state in which for a given input optical intensity the output intensity can be one of the two stable state. Bistable system can be realized by utilizing optical non-linearity and feedback. Here the output of the non-linear element is fed back to the device to control the transmission, absorption or other properties of the system so that bistability can be obtained. Implementation of optical bistable devices are based on.

1. Dispersive non-linear element—Refractive index or absorption is controlled as a function of light intensity.
2. Intrinsic feed back—Optical feed back required is applied internally.
3. Dissipative non-linear element—Absorption coefficient depends on the optical intensity.
4. Integrated hybrid devices—Optical bistability is achieved by integrating optical/electrical devices.

Figure 14.11 shows an optical bistable device implemented by an integrated directional coupler in which electric feed back is used. For the given input, the output is detected by a photo detection and fed back as voltage to the electrodes which modulate the refractive index of the waveguide. This modifies the coupling between the waveguide and the input, and act as a bistable device.

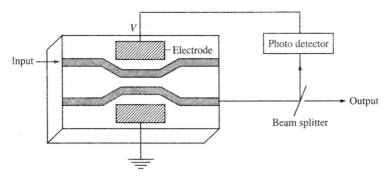

Figure 14.11 Integrated directional coupler acting as an optical bistable device.

Bistable devices are very important to implement optical switches, gates, flip-flop, etc., in optical computers.

14.9
OPTICAL INTERCONNECT

As interconnections between various components in electronic devices are achieved by conducting channels such as cables, wires, etc. photonic interconnections can be established by optical fibres, waveguides, integrated optic couplers, mirrors, lenses, etc. In addition to this free space guiding is also possible (which in not possible in electronic systems). Basic function of all these photonic interconnections is to guide light to different components of the system using appropriate optical interconnection via, shift, fan in and out, magnification, reduction, reversal, deflection, shuffle, etc. High density interconnections, many to one or one to many connection are possible by holographic interconnections. Active optical amplifier can also introduce which eliminate the need for repeators.

Optical interconnections offers high connectivity high speed, low loss and cross talks; parallelism makes it very attractive for optical computers. The basic advantages are as follows:

1. High speed data transfer at the speed of light.
2. Low power requirements; only limited by photodetectors, efficiency of electrical/optical or optical/electrical conversion transmission efficiency of the rooting elements.
3. Greater interconnection density can be achieved.
4. Delay time is very small as the data travels at the speed of light.
5. Optical interconnections have greater density-bandwidth product. Density is not effected by the bandwidth.
6. Electrical to optical or optical to electrical conversion can be achieved easily.

14.10
PHOTONIC SWITCHES

A photonic switch is a device (similar to an electrical switch) which establishes light on or off (or other physical parameters via light) condition between different components in a photonic device. Basic principle behind the photonic switches are discussed as follows:

Optomechanical switches

Optomechanical switches are based on the mechanical movements such as rotation, motion, etc. of mirrors, grating, prisms etc. by deflecting the beam path. Movements are usually done by using piezoelectric devices and the switch speed is only in range of milliseconds.

Electro optic switches

These switches are based on electro optic effect in which refractive index is controlled by applying an electric field. Their speed is very high (20 GHz) and operate at a few volts.

Magneto optic switches

Here the polarization is controlled by applying magnetic field. For example, for material exhibiting Faraday effect, the polarization rotates, the rotatory power is proportional to the magnetic field. Switching speed is 100 ns and more.

Acousto optic switches

Acousto optic switches are based on the Bragg deflection by sound. The power of the deflected light is controlled by the intensity of the sound while the deflection angle is controlled by the frequency of the sound.

Electronic switching

Here the optical signals are converted to electrical signal using photodetectors and then converted to light using LED or laser diodes. There is a large power loss, and delay in these switching devices.

Opto optic switches

In all optical switches, light controls light with the help of non-linear effects.

14.11
OPTICAL COMPUTERS

Optical computers make use of photons (electronic computer uses electrons) for implementing logic and arithmetic operations. Hardware are based on photonic switches, logic gates, memory devices; and optical interconnection based on holographic, waveguide structures and optical fibres, integrated couplers, lenses, mirrors, gratings, etc. are employed. The architecture of optical computer is similar to that of electronic digital computers with electronic hardware and connections are replaced by corresponding optical devices and interconnections. The logic states are represented by high or low light intensity levels corresponding to 1 or 0 states, which are attained by bright or dark intensity levels in the form of light pulses. Research on the development of optical computers is in progress by designing high performance optical hardwares and interconnections, and it is expected that optical computers with extremely high speed, small size, low power consumption, may replace conventional electronic computers in future.

14.12
ULTRAFAST PHENOMENA

Lasers with extremely short pulse duration are known as ultrashort lasers. For example, in fementosecond lasers, pulse widths are in the range of a few fementoseconds. Lasers with one attosecond pulse duration is now available. Since the pulse width is extremely small, the peak intensity is very large and can produce powder density larger than 10^{16} W/cm^2. In this intensity range, non-linear interactions of light with matter is considerably modified (as compared with the low power lasers). Recently with the advent of ultrashort pulse lasers, the research on ultrafast phenomena such as laser–matter interaction in intense field, higher order harmonics, laser plasma, super continuum etc. have great interest in physics (Super continuum–when an ultrashort pulse is passed through a medium like water a broad spectrum is generated, i.e., a monochromatic radiation generates a white light spectrum).

REVIEW QUESTIONS

1. Explain the principle of Raman effect.
2. Explain the principle of photorefractive effect.
3. What is Optogalvanic effect? Explain the experimental set up.
4. What is photothermal deflection?
5. Discribe the photo refraction in a diffusing medium.
6. What are squeezed states?
7. What are optical solitons?
8. Explain optical bistability. How it can be implemented?
9. What are optical interconnects?
10. Write a short note on optical computer.
11. How optical switches can be implemented?
12. Write a short note on ultrafast laser pulses.
13. What is POG effect?

References

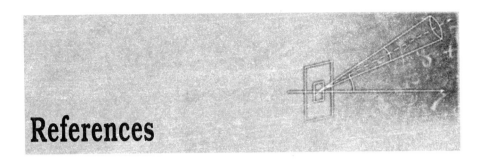

Bhattacharya, Pallab, *Semiconductor Optoelectronic Devies*, Prentice Hall of India, New Delhi, 1999.

Djafar, K. Mynbaev, Lowell L. Scheiner, *Fiber-optic Communications Technology*, Pearson Education, India, 2002.

Francis, T.S. Yu, Xiangyang Yang, *Introduction to Optical Engineering*, Cambridge University Press, UK, 1997.

Ghatak, Ajoy and K. Thyagarajan, *Optical Electronics*, Cambridge University Press, UK, 2003.

Gupta, S.C., *Textbook on Optical Fiber Communication and its Applications*, Prentice Hall, India, 2004.

Palais, C. Joseph, *Fiber Optic Communications*, PH International, NJ, 1998.

Powers, John, *Fiber Optic Systems*, McGraw-Hill, Singapore, 1999.

Singh, Jasprit, *Opto Electronics*, McGraw Hill, Singapore, 1996.

Sirohi, Rajpal S., *Wave Optics and its Applications*, Orient Longman, India, 1993.

Index

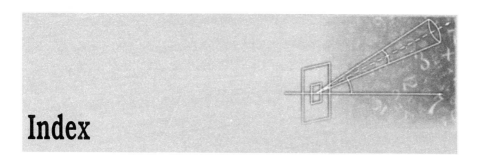

A, B coefficients, 176
Absorption, 121
    coefficient, 43
    direct band, 43
    donor-acceptor, 23
    free carrier, 22
    impurity band, 24
    indirect transition, 23
    intraband transition, 23
    low energy, 22
Acceptance angle, 102
Acceptor, 14
    level, 14
Acousto optic effect, 130
Active waveguide devices, 150
Amplitude modulation, 128
Anti-Stokes line, 193
Application of
    holography, 165
    laser, 190

Band structure, 16
Bandgap energy, 59
Bending loss, 110
Boltzmann distribution law, 11
Bragg
    diffraction, 131
    modulator, 153
Branching waveguide switch, 155
BSF, 93

Candela, 3
Carrier concentration, 13

Cathode ray tube, 167
CCD, 172
Characteristics of
    a hologram, 164
    laser, 174
CO_2 laser, 187
Coherent bundle state, 105
Colour holography, 163
Conduction band, 14
Coupler, 114
Coupling loss, 110
Critical angle, 119

Dark current, 59
    resistance, 60
Density of states, 12
    function, 12
DH, 45
    LED, 46
Differential quantum efficiency, 52
Diffuse reflector, 6
Direct band gap, 16
Directional coupler switch, 151
Dispersion, 200
Distributed feedback laser, 56
Donor, 10
    level, 14
Double-crucible method, 114
Dye laser, 188
Dynode, 75

Edge emitting LED, burrus LED, 46
Effect of doping, 13

Efficiency of LED, 142
Electron hole pair, 18
Electrooptic effect, 126
Exciton, 24
    auger recombination, 29
    Franz-Keldysh effect, 28
    SRH recombination, 29
    Stark effect, 28
    Stokes shift, 28
Extended source, 6
External modulation, 122

Fabrication of optical fibre, 142
Fabry-Perot cavity, 51
Fermi level, 10
Fermi-Dirac distribution function, 10
Fibre drawing process, 115
Fill factor, 88
Fluorescence, 34
Four level system, 183
    wave mixing, 139
Fractional change, 100
Frequency
    down conversion, 139
    up conversion, 139
Fresnel loss, 43

Glass fibre, 108
Graded index fibres, 110
Guided modes, 103, 150
    leaky modes, 103
    radiation modes, 103

Half wave voltage, 129
He–Ne laser, 185
Hetrojunction, 190
Hologram, 163
Holographic
    interferometry, 165
    memory, 165
    microscopy, 165
Homojunction, 89

Illuminance, 4
Indirect band gap, 17
Injection efficiency, 42

Integrated optics, 117
Internal quantum efficiency, 20
Intersystem crossing, 188
Intrinsic semiconductor, 13
Inverse square law, 5
Irradiance, 4

Johnson noise, 77

Kerr effect, 126

Lambert's law, 5
Laser action, 174
LED, 33
    drive circuit, 36
Light bending device, 145
Liquid crystal display, 167
Losses in a resonator, 182
Lumen, 2
Luminance, 4
Luminescence, 33
Luminosity curve, 1
Luminous
    emittance, 3
    energy, 2
    intensity, 3
    power, 2
Lux, 4

Macrobending, 113
Magneto optic effect, 125
Maximum power conversion efficiency, 88
Meridional rays, 101
Method of holography, 158
Microbending, 113
Microcavity photodiode, 73
MIE scattering, 111
Minority carrier injection, 34
Model dispersion, 105
Modulation of light, 122
MSM photodiode, 83
Multimode and single mode fibre, 104
Multiphoton absorption, 140
Multiplexer and demultiplexer, 117
Multiplication noise, 77

Nd: YAG laser, 184
Nematic liquid crystal, 170
Non-linear
    optical processees, 190
        Raman scattering, 193
Non-radiative recombination, 19
Numerical aperture, 101

Object wave, 56
Off axis holography, 161
Open circuit voltage, 94
Optical
    bistability, 156
    computer, 202
    fibre, 97
    parametric oscillations, 190
    power divider, 146
    resonator, 174
    solitons, 200
Opto
    electronic integrated circuit, 157
    galvanic effect, 196

Pauli's exclusion principle, 10
Perfect diffuser, 6
Phase modulation, 127
Phase
    conjugate waves, 140
    retardation, 128
Phosphorescence, 34
Photoconductive materials, 60
Photoconductor, 78
Photodetectors, 79
Photodiode
    design, 63
    equivalent circuit, 63
    expression for photocurrent, 65
Photometry, 1
Photomultiplier tube, 74
Photonic switches, 202
Photorefractive effect, 194
Photothermal deflection, 197
Phototransistor, 71
Photovoltaic effect, 84
PIN photodiode, 65
    avalanche multiplication factor, 68
    design, 69
    photocurrent, 69

Planar waveguide, 142
Planck's law, 176
Plasma display, 173
Plastic fibre, 108
Pockel effect, 126
Point source, 4
Polymer optical fibre, 108
Population inversion, 176
Power conversion efficiency, 52
Pulse dispersion, 109

Q factor, 181
Quantum well laser, 55
Quasi-Fermi level, 14

Radiance, 3
Radiant
    emittance, 2
    energy, 2
    intensity, 3
    power, 2
Radiative recombination, 19
Radiometry, 1
Raman active medium, 193
    effect, 193
Raman–Nath diffraction, 131
Rayleigh scattering, 111
Real image, 160
Recombination, 18
Reconstructing of hologram, 160
Recording of hologram, 160
Refractive index, 43
Reference wave, 160
Resonant modes, 53
Richardson–Dashman relation, 76
Ruby laser, 137

Scattering loss, 111
Schottky barrier photodiode, 71
Schottky barrier solar cell, 91
SH laser, 54
Squeezed states, 199
Step index fibre, 98
Stripe waveguide, 142
Surface acoustic waves, 153

Textured cell, 94
Thermal noise, 60
Thin film, 84
Threshold gain, 51

Ultrafast phenomena, 203

Valence band, 14
Volume hologram, 163
V-parameter, 103

Wavefront reconstruction, 159
Waveguide coupling, 159
   edge coupling, 156
   grating coupling, 157
   polarizer, 158
   prism coupling, 157
   structure, 142
White light hologram, 163
Workfunction, 74

Zero dispersion fibre, 109